W9-BZA-934

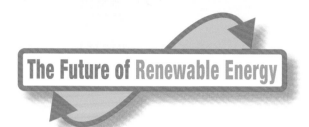

The Future of Renewable Energy

What Is the Future of Solar Power?

Andrea C. Nakaya

ReferencePoint
Press®

San Diego, CA

About the Author

Andrea C. Nakaya, a native of New Zealand, holds a BA in English and an MA in Communications from San Diego State University. She currently lives in Encinitas, California, with her husband and their two children, Natalie and Shane.

© 2013 ReferencePoint Press, Inc.
Printed in the United States

For more information, contact:
ReferencePoint Press, Inc.
PO Box 27779
San Diego, CA 92198
www. ReferencePointPress.com

Picture Credits:
Cover: Thinkstock.com
Thinkstock/iStockphoto.com: 8
Steve Zmina: 10, 16, 23, 30, 36, 43, 49, 56, 60

LIBRARY OF CONGRESS CATALOGING-IN-PUBLICATION DATA

Nakaya, Andrea C., 1976-
  What is the future of solar power? : part of the future of renewable energy series / by Andrea C. Nakaya.
      p. cm. -- (Future of renewable energy series)
  Includes bibliographical references and index.
  ISBN 978-1-60152-278-8 (hardback) -- ISBN 1-60152-278-9 (hardback) 1. Solar energy.
2. Photovoltaic power generation. I. Title.
  TJ810.N335 2013
  333.792'3--dc23
                                                           2012014162

Contents

Foreword

What are the long-term prospects for renewable energy?

In his 2011 State of the Union address, President Barack Obama set an ambitious goal for the United States: to generate 80 percent of its electricity from clean energy sources, including renewables such as wind, solar, biomass, and hydropower, by 2035. The president reaffirmed this goal in the March 2011 White House report *Blueprint for a Secure Energy Future*. The report emphasizes the president's view that continued advances in renewable energy are an essential piece of America's energy future. "Beyond our efforts to reduce our dependence on oil," the report states, "we must focus on expanding cleaner sources of electricity, including renewables like wind and solar, as well as clean coal, natural gas, and nuclear power—keeping America on the cutting edge of clean energy technology so that we can build a 21st century clean energy economy and win the future."

Obama's vision of America's energy future is not shared by all. Benjamin Zycher, a visiting scholar at the American Enterprise Institute, a conservative think tank, contends that policies aimed at shifting from conventional to renewable energy sources demonstrate a "disconnect between the rhetoric and the reality." In *Renewable Electricity Generation: Economic Analysis and Outlook* Zycher writes that renewables have inherent limitations that can be overcome only at a very high cost. He states: "Renewable electricity has only a small share of the market, and ongoing developments in the market for competitive fuels . . . make it likely that renewable electricity will continue to face severe constraints in terms of competitiveness for many years to come."

Is Obama's goal of 80 percent clean electricity by 2035 realistic? Expert opinions can be found on both sides of this question and on all of the other issues relating to the debate about what lies ahead for renewable energy. Driven by this reality, *The Future of Renewable Energy* series critically examines the long-term prospects for renewable energy by delving into the topics and opinions that dominate and inform renewable energy policy and debate. The series covers renewables such as solar, wind, biofuels, hydrogen, and hydropower and explores the issues of cost and affordability, impact on the environment, viability as a replacement for fossil fuels, and what role—if any—government should play in renewable energy development. Pointed questions (such as "Can Solar Power Ever Replace Fossil Fuels?" or "Should Government Play a Role in Developing Biofuels?") frame the discussion and support inquiry-based learning. The pro/con format of the series encourages critical analysis of the topics and opinions that shape the debate. Discussion in each book is supported by current and relevant facts and illustrations, quotes from experts, and real-world examples and anecdotes. Additionally, all titles include a list of useful facts, organizations to contact for further information, and other helpful sources for further reading and research.

Visions of the Future: Solar Power

Germany is a country about the same size as the US state of Montana and only receives about as much sunshine as the state of Alaska, yet it is a world leader in solar power. This relatively small country has installed more solar panels than any other country in the world and manages to generate a significant amount of solar power. By the end of 2011 it had almost half the world's photovoltaic (PV) solar power capacity. According to BSW-Solar, Germany's main solar industry association, in 2011 Germany produced enough solar power to supply 5.1 million households for a year. BSW-Solar estimates that this will only increase, predicting that between 2012 and 2016, the share of solar power in Germany's energy mix will grow by 70 percent. The organization says, "Solar power [is] on the path to becoming a key pillar of sustainable energy supply in Germany."[1]

Most of the world's energy is currently supplied by fossil fuels (coal, oil, and natural gas). However, many people believe that these fuels are not a good choice as a source of power because they generate significant amounts of pollution and are limited in quantity. Solar power is viewed by many as a cleaner and more sustainable alternative. The sun contains far more energy than fossil fuels. According to the Union of Concerned Scientists, "All the energy stored in Earth's reserves of coal, oil, and natural gas is matched by the energy from just 20 days of sunshine."[2] Germany proves that it is possible to harness that energy and make solar power an important part of a country's energy supply.

Capturing the Power of the Sun

Solar power involves harnessing the sun's energy to produce heat, lighting, or electricity. That energy is captured in a number of different ways. One of the most common technologies is solar PV. PV technology uses cells made of semiconductor materials, which convert sunlight directly into electricity. This technology is generally used on a smaller scale; for example, on the roof of a house or business to provide electricity for just that building. But it can also be used at utility scale, where a large power plant is constructed with many PV panels. In the past, PV technology was commonly used to provide power for areas that were not connected to the power grid, such as homes in remote locations. Increasingly, however, it is being used in connection with the power grid. According to energy expert Dan Chiras, the majority of new solar electric systems in the United States are connected to the grid.

Another common solar power technology is concentrating solar power (CSP). CSP uses mirrors or lenses to concentrate a large area of sunlight onto smaller receivers that collect this heat and turn it into power. CSP technology is used in large solar power plants, and the receivers used can be various shapes, such as flat panels or troughs. A major advantage of CSP is the potential for storing the energy collected and using it later; for example, when the sun is not shining at night.

Solar power is also commonly used for space heating and cooling in buildings and to heat water heaters and pools. These uses have become increasingly common in new homes and businesses. For example, in 2008 Hawaii passed a law requiring solar water heaters to be installed in new homes, the first state in the United States to do so.

Exponential Growth for Solar Power

Overall, the world generates a minimal amount of power—less than 1 percent—through solar technology. In recent years, however, exponential growth has made solar power one of the fastest-growing forms of renewable energy. While it is still more expensive than many other types of power, there have been continual and dramatic reductions in cost, and the technology continues to become more efficient and more widely

Rooftop solar panels (pictured) are just one of the technologies that can capture the power of the sun. Solar power is one of the fastest-growing forms of renewable energy but where it will fit in the world's energy picture is still uncertain.

used. According to the Solar Energy Industries Association, "As a result of growing awareness about reliable . . . solar technology, concerns about rising costs of conventional energy, and new state and federal incentives, deployment of solar energy has exploded since 2005."[3] In 2011 the International Energy Agency estimated that with international commitment to solar power, solar technologies could provide a third of the world's energy by 2060.

In the United States, solar power supplies less than 1 percent of the electricity used; however, it is a fast-growing technology. According to the Solar Energy Industries Association, solar is the fastest-growing energy sector in the United States. The association predicts that by 2014, solar will be the largest source of new electric capacity in the United States.

Growth Trends

The majority of solar power installed around the world is PV power. According to the European Photovoltaic Industry Association's *Market Report 2011*, at the end of 2011 the total worldwide capacity of PV connected to the existing power grid was 67.4 gigawatts (GW). Gigawatts (and megawatts) are common ways to measure electrical power. The capacity is the amount of power that can theoretically be generated with existing installations; however, actual generation may be lower, depending on factors such as weather conditions and equipment problems.

Much of the world's existing capacity was added in recent years: 16.6 GW in 2010 and 27.7 GW in 2011. The European Photovoltaic Industry Association says that the majority of the growth in 2011 occurred in Europe, where Italy installed 9 GW and Germany installed 7.5 GW. In addition, China installed 2 GW and the United States 1.6 megawatts (MW). Overall, the world leader in the total amount of PV power installed is Germany, followed by Italy, then Japan.

> "Deployment of solar energy has exploded since 2005."[3]
>
> —The Solar Energy Industries Association, the US trade association for solar energy.

The European Photovoltaic Industry Association predicts a strong future for PV solar power. It says:

> PV is now, after hydro and wind power, the third most important renewable energy in terms of globally installed capacity. The growth rate of PV during 2011 reached almost 70%, an outstanding level among all renewable technologies. The total energy output of the world's PV capacity run over a calendar year is equal to some 80 billion kWh [kilowatt-hours, a measurement of electrical power]. This energy volume is sufficient to cover the annual power supply needs of over 20 million households in the world.[4]

While CSP installations are far less prevalent than PV power, CSP has also experienced substantial growth in recent years, and experts believe it will play a significant role in the future energy supply. In a 2011 report,

US Solar Power Potential

The United States has enormous solar power potential. The region with the greatest potential is the southwest where hot, sunny weather is common.

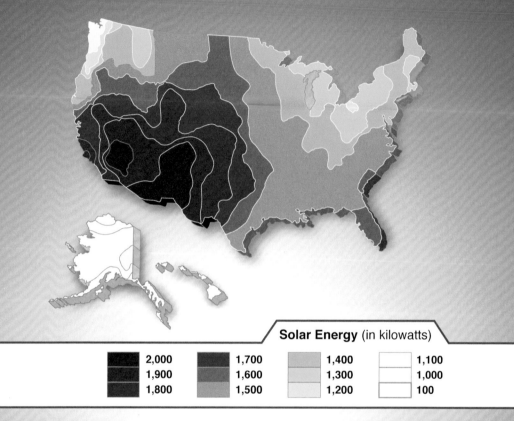

Solar Energy (in kilowatts)

■ 2,000	■ 1,700	■ 1,400	□ 1,100
■ 1,900	■ 1,600	■ 1,300	□ 1,000
■ 1,800	■ 1,500	■ 1,200	□ 100

Source: Solar Energy Industries Association and Prometheus Institute, "US Solar Industry: 2007 Year in Review," 2008. www.seia.org.

the Renewable Energy Policy Network for the 21st Century (REN21) finds that between 2007 and 2010, about 0.74 GW was added world-wide, more than half of it in 2010. REN21 estimates that at the end of 2010 the global total was 1.095 GW. Of that total, Spain and the United States dominate. REN21 reports: "CSP growth is expected to continue at a rapid pace."[5] In 2011 the organization found that both the United States and Spain had substantial new CSP installations under construction. However, in comparison with PV installations, CSP numbers are small. While there was 1,095 MW of CSP globally at the end of 2010, REN21 finds that PV was about 40,000 MW.

Obstacles to Development

Solar power has enormous potential because of how much energy the earth receives from the sun. Harnessing that energy in ways that will be useful to society—and affordable—represents a formidable challenge, however. Affordability is the biggest obstacle to widespread use of solar power. Solar technology has undergone dramatic price reductions and prices continue to fall, but solar power is still a relatively expensive form of energy. In addition, many people argue that even if solar power were affordable, it is simply not reliable or efficient enough to meet society's energy needs. Critics contend that it cannot survive in the energy market without substantial and continued government support. Even with substantial recent growth, solar power only makes up a small share of the energy produced in the United States and the world, and it has a long way to go before it produces a major part of the energy supply.

A Strong Future

Despite these challenges, the use of solar power has spread rapidly around the world in recent years, and this growth is expected to continue. Researchers at Deloitte Consulting believe that solar power will have an important place in the future energy supply as the world is forced to utilize new energy sources. The researchers say, "In the coming decades, the world will likely confront a new energy crisis: combining rapid demand growth and strained supply with increased environmental and independence concerns. The rise of renewable energy sources is inevitable, and solar is particularly well suited for rapid growth as a result of its abundance and broad availability."[6] In its *Annual Energy Outlook 2012*, the US Energy Information Administration (EIA) predicts that between 2010 and 2035, consumption of solar energy in the United States will grow the fastest of all types of renewable energy. This powerful and inexhaustible resource is likely to play an important role in fulfilling the world's future energy needs.

Chapter One

Is Solar Power Affordable?

Solar Power Is Affordable

Solar power is experiencing dramatic price reductions and quickly approaching the point that it will cost the same as other types of power. It already saves consumers money in the long term, since after the initial investment is paid off there are no ongoing fuel costs. As fossil fuel prices increase in the future, solar power will become even more affordable by comparison. Solar power's affordability is further enhanced by the large number of jobs it generates. Installing solar power is a practical and cost-effective option.

The Debate

Solar Power Is Too Costly

Compared to other ways of generating power, solar power is too costly, and its economic benefits have been overstated. Even with government financial incentives, solar energy is still significantly more expensive than natural gas and coal, and even many other renewable sources such as wind and hydropower. For the average household a PV system is too expensive to purchase and takes too long to pay off. Even with continuing price declines in solar power, this price difference is expected to remain.

Solar Power Is Affordable

"Solar is now cost-effective."

—Paul Krugman, winner of the Nobel Prize in Economics in 2008 and a *New York Times* columnist.

Paul Krugman, "Here Comes the Sun," *New York Times*, November 6, 2011. www.nytimes.com.

The common belief that solar power is too expensive is incorrect. In reality, solar power is an affordable form of power, and it is nearing the point that it will cost the same as traditional sources of energy. Recent dramatic price declines make solar power—especially PV technology—affordable. According to Rhone Resch, president and CEO of the Solar Energy Industries Association, in 2010 alone, the price of solar energy fell 20 percent in the United States.

Public perceptions have not kept pace with reality, say researchers at Queen's University in Canada. Their 2011 study of PV power found that most cost estimates for solar power are outdated. These unrealistic estimates ignore recent dramatic reductions in the cost of solar panels and underestimate both the lifetime and productivity of these panels. Joshua Pearce, one of the study's authors, believes that PV is actually near the point that it can produce energy for about the same price as traditional sources of energy. CSP technology, while a little more expensive than PV, is also an affordable way to generate power. Experts estimate that solar power will achieve grid parity in the next decade or two, meaning that the cost will be the same as the cost of conventional sources of power from the power grid. The International Energy Agency predicts that grid parity will be achieved by 2020. In the United States, the Department of Energy (DOE) aims to achieve grid parity by 2015.

Prices Continue to Decrease

The price of solar power has decreased dramatically in the past few years, and this trend is expected to continue as solar technology improves.

According to the Pew Center on Global Climate Change, when solar PV technology was first developed in the 1950s, it cost $300 per watt of electricity produced. By 1998 the cost had dropped to $10.80 per watt, and by 2009 it had decreased to $7.50 per watt. *New York Times* columnist and Nobel Prize–winning economist Paul Krugman argues, "If the downward trend [in price] continues—and if anything it seems to be accelerating—we're just a few years from the point at which electricity from solar panels becomes cheaper than electricity generated by burning coal."[7] Mark M. Little, the global research director for the General Electric Company, predicts that as a result of technological improvement and continued price decreases, solar power may be cheaper than electricity generated by fossil fuels and nuclear reactors within 3 to 5 years.

Fossil fuels currently provide most of the world's energy, but the price of these fuels will increase in the future, due to decreasing supplies. In contrast, the price of solar power is likely to decrease, so in comparison to fossil fuels, solar power will become more affordable. Experts disagree over what amount of fossil fuels are left in the earth and how long this supply will last. However, most people agree that fossil fuels are a finite resource and therefore will become increasingly scarce over time. As with most any resource or product, scarcity usually leads to higher prices. Steven Cohen, executive director of Columbia University's Earth Institute, says, "The cost comparison between fossil fuels and solar is not a question of 'if' but one of 'when.' The cost curve [for solar power] is clearly headed down. While fossil fuels are still plentiful and relatively inexpensive, they are finite resources that over time will only become more expensive."[8]

Saves Money Long-Term

Solar power does require a significant initial investment for solar panels or other devices, but after that investment is paid off, solar systems save money because, unlike conventional power stations, they have no fuel costs. Estimates vary widely on the time it takes to recover the initial costs and start making a profit. Once it reaches that point, solar technology will provide many years of inexpensive power. In a 2010 article in *Popular*

Mechanics, journalist Elizabeth Svoboda says, "Solar panels are certainly expensive . . . but eventually, you're destined to end up on the positive side of the equation."[9] Svoboda argues that even if it takes 15 years to break even on a solar system—a conservative estimate—the system will still end up saving money eventually. She says, "Assuming solar cells have an average life expectancy of 30 years, more than 50 percent of the power solar cells generate ends up being free."[10]

Because solar systems have no fuel costs, the cost of power will not fluctuate, as fossil-fueled power rates do when fuel prices change. Krystal Book of American Solar Electric says, "When utility power rates increase, your solar power stays exactly the same." According to Book, "Most indicators suggest that utility rates will increase 5% annually for years to come. Not many home improvements allow you to decrease your cost of living. Solar offers you a hedge against rising utility rates."[11]

> "If the downward trend [in price] continues—and if anything it seems to be accelerating—we're just a few years from the point at which electricity from solar panels becomes cheaper than electricity generated by burning coal."[7]
>
> —Paul Krugman, economist, *New York Times* columnist, and winner of a Nobel Prize in Economics.

Even if the price of a home or business PV system is too high, there are still options that can make solar power affordable. In addition to numerous government grants and rebates, power companies are beginning to work with individuals and businesses to make solar systems more attainable. David Crane, CEO of NRG Energy, one of the largest electric companies in the United States, admits that a solar installation is affordable for very few people, so as a result, "what the industry is already fast creating is lease arrangements, and power purchase arrangements."[12] With these types of arrangements, the solar provider pays for the solar system, and the building owner pays a set fee for the use of the equipment or the power generated from it. With options such as these, it is not necessary to have the money to cover the whole cost of a system up front.

The Cost of Solar Power Continues to Decline

The history of solar power prices shows continuing and dramatic declines, making this technology increasingly affordable. This graph shows the price of photovoltaic (PV) solar cells per watt of power produced, between 1985 and 2011. (Prices are given in 2011 dollars.) In that period of time, PV cells have decreased from almost $7 per watt to less than $2 per watt. Similar price reductions are expected in the future as technology continues to improve.

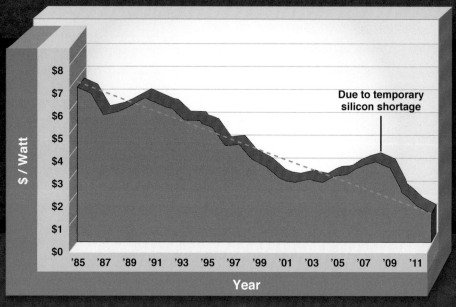

Source: Richard W. Caperton, "Good Government Investments in Renewable Energy," *American Progress*, January 10, 2012. www.americanprogress.org.

Benefits the Economy Through Job Creation

Solar power's affordability is enhanced by the way it helps the economy through job creation. The solar industry employs people in research and design of components and systems and in manufacturing, installation, and operations. In a 2012 visit to a solar facility in Boulder City, Nevada, President Barack Obama reported that the United States has more than 5,600 solar companies. These companies are providing a significant number of new jobs. According to Solar Foundation executive director Andrea Luecke, "Solar energy is a smart investment. It is clean, reliable energy that creates jobs. It is creating jobs at an impressive clip."[13]

Research shows that the solar industry employs more than 100,000 people in the United States. Luecke reports that jobs are being added much faster in the solar industry than in other parts of the economy. She reports that between 2010 and 2011, jobs in the solar industry grew 6.8 percent, compared to overall US job growth of only 0.7 percent in that period. In an article in *Energy Policy*, researchers found that renewable energy sources such as solar power are better for the economy than fossil fuels. The researchers reviewed 15 studies about the job creation potential of the clean energy industry and found that the clean energy sector generates more jobs per unit of energy delivered than the fossil fuel energy sector.

> "Solar energy is a smart investment. It is clean, reliable energy that creates jobs. It is creating jobs at an impressive clip."[13]
>
> —Andrea Luecke, executive director of the Solar Foundation, a nonprofit organization that works to increase the use of solar energy around the world.

An Affordable Option for Power

Many discussions of solar energy's affordability take place in the abstract. Tom and Elizabeth Coates, who operate Aunt Betsy's Chicken Farm near Detroit, Michigan, provide a concrete example of the financial benefits of solar power. The Coateses have found innovative ways to collect and use solar power, and those efforts have brought impressive results. They produce about 95 percent of their own electricity with solar power, an achievement that has proved to be not just affordable, but far cheaper than conventional energy. Before they started using solar power, they could not afford to expand the farm because the cost of electricity was too high. Elizabeth says, "To operate the farm and the house with regular electricity supply it would cost $800 a month. That was with no air conditioners, one freezer and one refrigerator, and I had a tenth of the farm we now have." With solar energy providing their electricity, she says, "we have the house, two portable freezers in trailers, five freezers, three refrigerators and air conditioning. We are also running incubators and have six fans in the barns [when it is hot]. . . . In the winter we run heat lamps and heaters."[14] Yet even though the farm's electricity needs are far greater now, with solar power Elizabeth says their electricity costs less than $100 a month.

While prices may still be higher than other forms of energy, overall, solar is a very affordable form of power. And when its numerous other benefits are included, such as the value of the jobs it creates and the costs saved by reducing environmental harm, solar power is a better choice for society. Researchers Richard Perez, Ken Zweibel, and Thomas E. Hoff maintain, "Society gains back the extra we pay for solar. It gains it back in a healthier, more sustainable world, economically, environmentally, and in terms of energy security."[15]

Solar Power Is Too Costly

"Solar power is still expensive relative to other forms of electricity."

—The Pew Center on Global Climate Change, now known as the Center for Climate and Energy Solutions, a nonprofit organization that works to advance policy and actions that address the problems of energy and climate change.

Pew Center on Global Climate Change, "Solar Power," Climate Techbook, August 2011. www.pewclimate.org.

According to a 2011 survey of 1,000 Americans by the polling firm Kelton Research, 9 out of 10 people believe it is important for the United States to develop and use solar power. Yet while the majority of the population supports the use of solar power, very few people actually have solar panels on their houses. This is because solar power is too expensive for most people.

Solar power is a far more costly way to produce power than other energy sources, such as coal, natural gas, and even many other renewable energy sources. In comparison to these less costly options, it is not an affordable choice for energy production. Estimates from the EIA illustrate just how expensive solar power is. In a 2010 report, the EIA estimates levelized costs for electricity generation for plants beginning service in 2016. Levelized cost is a common way to compare electricity costs. It is calculated by taking the building and operating costs for a power plant over its assumed lifetime, then using that total to calculate the average costs per megawatt-hour (MWh) of electricity produced. The EIA estimates that for 2016 plants, the average levelized cost of conventional coal will be $94.80 per MWh, and conventional natural gas will be $66.10 per MWh. Solar PV technology is estimated to be $210.10 per MWh, and CSP is $311.80 per MWh. So according to these estimates, solar power costs more than twice as much as conventional forms of energy. Solar is also more expensive than nuclear power and many renewable technologies. In fact, of all the energy technologies compared in the EIA report, the only power more expensive than solar is offshore wind, and that is only more expensive than solar PV. Solar thermal is ranked as the most expensive of the options compared.

Too Costly Without Government Support

Solar power only appears to be affordable because it receives large government loans, tax breaks, and other financial incentives. For example, according to a 2011 *USA Today* article by journalist Wendy Koch, before it ended its loan guarantee program late in 2011, the US Department of Energy gave up to $13.6 billion to 17 large solar projects. Without this help, however, the solar industry proves far too costly as a way to generate power. David Bergeron, president of SunDanzer Development, discusses PV technology. He argues, "Solar Photovoltaic (PV) electric panels are far too expensive to provide a sustainable energy alternative to homes and businesses already connected to the electric utility grid. The solar industry and solar jobs are artificial and only exist because of large government subsidies." He predicts, "When the subsidies end, the solar bubble will burst and most of the jobs and industry will vanish overnight. This is because the underlying economics of Solar PV are not viable."[16]

Koch cites energy economist David Kreutzer, who says that the solar market in Europe has already started to collapse due to diminishing subsidies and predicts a similar situation for the United States in the future. According to Mark Bachman, a renewable energy analyst at Avian Securities, in the next few years only 20 to 40 of the few hundred solar panel makers around the world are expected to still be in business.

> "When the subsidies end, the solar bubble will burst and most of the jobs and industry will vanish overnight. This is because the underlying economics of Solar PV are not viable."[16]
>
> —David Bergeron, president of SunDanzer Development, a solar energy company located in Tucson, Arizona.

Not Affordable for Most People

Even with existing tax breaks and other incentives, the price of a PV system is too expensive for most families. Estimates of the price of a solar system for a single family residential home vary by state, but overall the required investment is substantial. According to a 2011 *Forbes* article, the average total system cost is $35,967. Even with a federal tax credit, the average homeowner can expect to pay $20,892. Says the *Deloitte Review*, "Purchasing a solar system to power the average American home currently costs about as much as buying a new car. Even when prices fall, the investment will remain a stumbling block for most households."[17] Eventually—when the purchase and installation costs are paid—a solar system provides free power; however, this does not happen for many years. Estimates on how long it takes to pay back the cost of a system also vary greatly, but most people estimate it takes 10 to 15 years.

CSP plants also need government support in order to be constructed. These plants cost millions of dollars to build, and according to the EIA, the electricity they produce is far more expensive than any other source. So existing CSP power plants are only viable because of significant government support. If that support did not exist, it is unlikely that the CSP plants would either, because they are such a costly method of power generation.

Far from Being Competitive

Declining prices for solar panels have lured some into thinking that solar power will soon be affordable. But even with these lowered costs, the overall cost of installing a solar power system remains high, and this is not likely to change anytime in the near future. Bergeron notes that the total installed cost of solar panels would need to be $1 per watt or less to be economically feasible. The problem is, says Bergeron, "solar panels are about half the total system cost. The remaining cost is mounting hardware for the panels, the inverter to make AC power, wiring, labor, and permitting. Therefore, even if it were possible to manufacture panels free, the balance of system cost is still about $2 per Watt and the industry would continue to be non-sustainable without substantial subsidy."[18]

21

BP, one of the largest energy companies in the world, recently announced that it is withdrawing from the solar business because the industry is not profitable enough. The company entered the solar industry early on and had significant solar interests around the world. In a 2002 press release, BP reported that it had nearly 20 percent of the global solar market share and approximately $300 million in revenue. However, in 2008 it began to scale back its solar operations. Explained Tony Hayward, BP's chief executive in 2009, "[Solar power] is not going to make the transition to be competitive with more conventional power, the gap is too big."[19] Today, BP says that it is pursuing more profitable types of renewable energy. "BP Alternative Energy is focusing on those sectors of the energy industry where we can profitably grow our business," the company says. "This has led us to focus on wind and biofuels, businesses that are material, scalable, and suited to BP's core capabilities. . . . In keeping with this strategic focus, BP is winding down its solar operations."[20]

Not Economically Beneficial

Solar power does not actually benefit the United States by boosting the local economy and creating jobs. Rather, it simply takes money and jobs away from other parts of the economy. While studies do show that the solar industry results in new spending and jobs, this comes at the expense of other economic sectors. Bergeron says, "Perhaps the most egregious myth is the claim we are helping our economy and creating jobs. This is false. Money for the solar subsidies comes from taxpayers and ratepayers." While this spending benefits the solar industry, it may be harmful to other sectors. Bergeron adds, "As the money is taken from us, spending for other goods and services must fall. This causes economic and job losses in other segments of the economy, such as in restaurants, stores, and service and manufacturing companies."[21]

New York Times journalist Matthew L. Wald agrees that the creation of solar jobs does not mean an overall increase in the number of jobs available, but instead means the loss of other jobs in the economy. He says, "Build enough solar plants and some coal plants will shut down; that would amount to firing Peter to hire Paul."[22]

Solar Power Cost Exceeds Other Forms of Electricity Generation

Solar power is a more expensive way to generate electricity than just about every other power source. Offshore wind power is the only power source that rivals solar in terms of expense. This graph compares the costs of electricity generation for various energy sources. Costs are in dollars per megawatt-hour of electricity produced, a common way to measure electrical power.

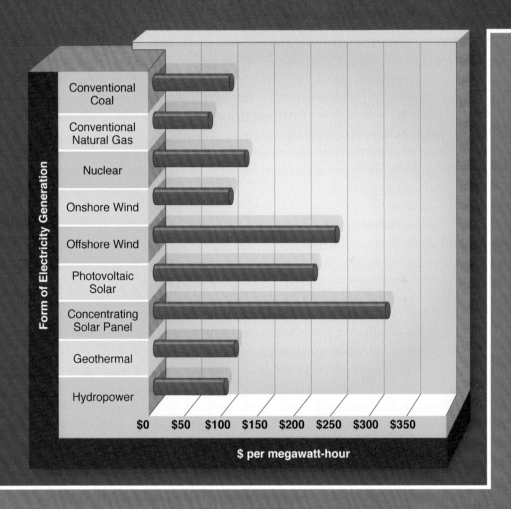

Form of Electricity Generation

- Conventional Coal
- Conventional Natural Gas
- Nuclear
- Onshore Wind
- Offshore Wind
- Photovoltaic Solar
- Concentrating Solar Panel
- Geothermal
- Hydropower

$0 $50 $100 $150 $200 $250 $300 $350

$ per megawatt-hour

Source: Institute for Energy Research, "Levelized Cost of New Electricity Generating Technologies," February 1, 2011. www.instituteforenergyresearch.com.

In addition to the fact that solar jobs displace other jobs, studies show that solar power is a much more expensive way to create jobs than other industries. Vance Ginn, former energy policy fellow at the Texas Public Policy Foundation, examines some recent solar power loans from the DOE's Loan Programs Office. He compares the total loan amount to the estimated number of jobs to be created from each solar project and finds that creating these jobs is very expensive. For example, says Ginn, "A previous guarantee for a solar panel project was $1.6 billion for BrightSource Energy, which produced 1,000 temporary construction jobs and only 86 permanent jobs. The cost of these permanent jobs was $18.6 million per job created."[23]

> "Solar power remains expensive relative to electricity produced using traditional fossil fuel generation sources, as well as certain renewable energy sources like wind."[24]
>
> —The Center for Climate and Energy Solutions, a nonprofit organization that works to address the challenges of energy and climate change.

While there have been recent dramatic cost reductions, solar power is still expensive compared to the alternatives. In an analysis of solar power, the Center for Climate and Energy Solutions maintains, "Solar power remains expensive relative to electricity produced using traditional fossil fuel generation sources, as well as certain renewable energy sources like wind."[24] Its high cost, and the availability of so many lower-cost alternatives, makes solar power a poor choice for widespread use.

Chapter Two

How Does Solar Power Impact the Environment?

Solar Power Is Beneficial to the Environment

Solar power is environmentally beneficial, especially when compared to the coal and natural gas that society currently uses for most of its energy. Both of these fossil fuels cause significant environmental destruction and harm human and animal health. In contrast, solar power has very few environmental impacts. Current solar power projects will reduce greenhouse gas emissions and other types of pollution and provide power for society with far less harm to the environment.

The Debate

Solar Power Does Not Benefit the Environment

Solar power actually has a number of negative impacts on the environment. To create a meaningful amount of solar power requires vast amounts of land, causing harm to the natural environment and displacing both people and animals. Solar power can also use substantial amounts of water and cause pollution during both the manufacturing and disposal of components. Existing solar power projects illustrate the reality of these many environmental harms.

Solar Power Is Beneficial to the Environment

"Solar energy is one of the most environmentally friendly renewable energy sources, and using more solar energy would bring great environmental benefits."

—Ned Haluzan, author of numerous renewable energy articles and the blog *Renewable Energy Articles*.

Ned Haluzan, "Why Is Solar Energy Good for the Environment?," *Renewable Energy Articles* (blog), November 23, 2010. www.renewables-info.com.

The production of solar power results in far less environmental destruction and pollution than conventional sources of power, such as coal, natural gas, and nuclear power. Using solar power instead of these fuels would significantly benefit the environment. Coal is the main source of electricity in the United States and an important energy source around the world, but it causes numerous environmental harms. For example, coal mining destroys the land and can pollute the water, harming both human and animal health. According to the EIA, about 70 percent of the coal mined in the United States comes from surface (or strip) mines, where miners remove the rock and soil above coal deposits. This type of mining disturbs the surface of the land and destroys the natural habitat. The EIA says:

> One surface mining technique that has affected large areas of the Appalachian Mountains in West Virginia and Kentucky is mountain top removal and valley fill mining, where the tops of mountains have been removed using a combination of explosives and mining equipment and deposited into nearby valleys. As a result, the landscape is changed, and streams may be covered

with a mixture of rock and dirt. The water draining from these filled valleys may contain pollutants that can harm aquatic wildlife downstream.[25]

In addition, coal results in high levels of pollution when it is burned to create power; that pollution can cause lung and heart damage and trigger asthma and other breathing problems. According to the Sierra Club, pollution from coal burning can be blamed for 12,000 emergency room visits every year in the United States and $100 billion in health care costs.

Natural Gas

Natural gas is another common source of power. Used for heating, cooking, electricity, and transportation, it provides about a quarter of the energy in the United States. Like coal, natural gas production involves significant disruption of the landscape. The Union of Concerned Scientists says, "Like most energy sources, natural gas production inevitably disturbs the natural landscape, as land must be cleared for roads, drilling equipment, processing facilities, and pipelines. Gas exploration, drilling, and production can create traffic congestion, road damage, dust, and noise in local communities."[26] Although the natural gas industry contends that new drilling technology has reduced the surface disruption associated with recovering natural gas, the Union of Concerned Scientists disagrees. It argues that even though the new technology causes less disruption, the gas industry is still causing significant environmental harm because new natural gas discoveries are frequently in undeveloped areas. In these areas any development destroys the natural landscape.

In addition to environmental disturbance, natural gas can also cause pollution. While natural gas burns much more cleanly than coal does, producing that gas can cause pollution. Journalist Elizabeth Shogren writes about the Marcellus natural gas deposit in the eastern United States. She says, "As companies race to produce gas from the enormous formation, they're operating thousands of new pollution sources. Compressor stations, drill rigs, processing plants, pipelines, diesel trucks and other equipment already leak pollution across large stretches of West Virginia and Pennsylvania."[27] Shogren relates the story of the Judy family, which lives near numerous facilities that compress natural gas for a pipeline. Since the compressor stations were put in, the family has been experiencing headaches, fatigue, dizziness, nausea, and nosebleeds. When state officials tested the air near the family's house, they found numerous toxic chemicals, which they say most likely came from the compressor stations.

Reducing Greenhouse Gases

Because the production of solar power does not generate any pollutants, it can also reduce the world's emissions of greenhouse gases. Greenhouse gases are a mixture of carbon dioxide and other pollutants that collect in the atmosphere and trap the sun's heat. Burning fossil fuels such as coal and natural gas is a major cause of greenhouse gas. Climate scientists say that the steady buildup of greenhouse gases over the last several decades is causing an increase in global temperatures that will have harmful effects on the environment and human health. Replacing greenhouse gas–emitting fuels such as coal and natural gas with solar energy could significantly reduce these emissions. According to the Natural Resources Defense Council, the United States is the world's second-largest contributor to greenhouse gas emissions, after China, and one of the largest sources of emissions worldwide is coal-burning power plants. So, greater use of solar power and less reliance on coal power could have a significant impact on greenhouse gas emissions. The Center for Climate and Energy Solutions says that by 2050, solar PV around the world could avoid 3.2 percent of global emissions from the electricity sector.

The California Valley Solar Ranch is one example of how greenhouse gases can be reduced through solar power. When complete, the power plant is expected to power 100,000 homes, and its developer, SunPower, says this will help cut down on greenhouse gas emissions. According to SunPower, the solar ranch's power production will offset the production of more than 750 million pounds (340 million kg) of carbon dioxide per year, the equivalent of taking 23 billion pounds (10.4 billion kg) of coal out of power plant production.

Minimal Environmental Impact

Overall, the environmental impact of solar power is far less harmful than that of the fossil-fueled energy sources society currently relies on. Using solar power instead of these fuels would be beneficial because it would reduce the overall environmental impact of generating power. Says Monique Hanis, spokesperson for the Solar Energy Industries Association, "We need to get our energy from somewhere, and solar energy is a cleaner, safer option than the status quo."[28]

> "We need to get our energy from somewhere, and solar energy is a cleaner, safer option than the status quo."[28]
>
> —Monique Hanis, a spokesperson for the Solar Energy Industries Association, a US solar energy trade association.

Manufacturing the components of a solar system, such as PV panels, does result in some pollution; however, the overall pollution generated is minimal. Mona Reese is cofounder of Brightstar Solar, a New England–based company that markets, designs, and installs PV systems. She explains, "You unfortunately have to consume a little energy to save a lot more. After a [PV] module is manufactured and installed, it will be a zero emissions energy source for the rest of its life. Over a 30 year life expectancy, almost 90% of the energy generated from the solar panel will be pollution free."[29] According to the Center for Climate and Energy Solutions, it currently takes about 4 years to recover the energy used to produce an average PV system, meaning that after 4 years the system begins to save energy overall. The center believes that technological improvements will bring this down to only 1 or 2 years. In contrast, coal power plants consume energy and produce pollution over their entire lifetime.

Solar Power Will Reduce Illness from Fossil Fuel Pollution

Pollution resulting from the operation of fossil-fueled power plants causes thousands of deaths and illnesses every year in the United States, according to research findings that appear in *Scientific American*. This chart shows the most common health problems, and the mean, or average, number of cases that occur each year. In contrast, the production of solar power causes almost no pollution and will thus significantly reduce the current health burden caused by fossil fuels.

| US Health Burden Caused by Particulate Pollution from Fossil-Fueled Power Plants ||
Burden	Number of Cases per Year
Pneumonia (hospital admissions)	4,040
Cardiovascular ills (hospital admissions)	9,720
Premature deaths	30,100
Acute bronchitis cases	59,000
Asthma attacks	603,000
Lost workdays	5,130,000

Source: Mark Fischetti, "The Human Cost of Energy," *Scientific American*, September 2011. www.scientificamerican.com.

Some critics of solar power argue that it uses large amounts of water. CSP power plants do require water for cooling and cleaning; however, that use is reasonable compared to society's many other water uses. New and continuing technological improvements in CSP plants have greatly reduced the amount of water needed for solar power generation. Hanis explains, "Many solar power plants are employing dry cooling technologies to reduce their water needs by as much as 95 percent."[30] She points out that society uses significant amounts of water to satisfy its other needs, such as growing food, and suggests that it is reasonable to also use part of the water supply for the purpose of creating power. She says, "Water use must be put in context: Independent studies by the US Geological Survey and the National Renewable Energy Laboratory show that solar facilities use about a quarter of the water per acre that agriculture uses."[31]

Better than the Other Alternatives

The Palen Solar Power Plant in Riverside, California, is an example of how solar power can benefit the environment. The plant is slated to begin construction in 2014 and is intended to provide significant environmental benefits with a minimal impact on the environment. It will provide power for approximately 150,000 homes. According to project developers, the power plant will have numerous environmental benefits, including reducing carbon dioxide emissions by almost 442,000 tons (400,976 metric tons) per year, nitrogen dioxide by 340 tons (308 metric tons), and sulphur dioxide by 290 tons (263 metric tons). In addition, Solar Trust of America, the project developer, maintains that the company is going to great lengths to minimize any environmental harm. Solar Trust says that it aims to have as little impact as possible on the landscape. Furthermore, says the

> "Over a 30 year life expectancy, almost 90% of the energy generated from the solar panel will be pollution free."[29]
>
> —Mona Reese, cofounder of Brightstar Solar, a New England–based company that markets, designs, and installs PV systems.

company, "in response to comments regarding habitat, Solar Trust of America has purchased thousands of acres of private land that will serve in perpetuity as high-quality habitat for any species that require relocation."[32]

Even small solar systems, such as PV systems designed to provide power for a single house, are beneficial to the environment. According to solar company SolarCity, over 20 years a solar system for a typical three-bedroom home will offset approximately 100,000 pounds (45,360kg) of carbon dioxide—similar to the amount of carbon dioxide created by driving a car 100,000 miles (160,934km).

Any type of power production has some environmental impacts, and solar power is no exception. However, compared to the alternatives, the impact of solar power is relatively small. The fossil fuels that society currently uses for most of its power inflict many environmental harms. Steven Cohen, executive director of Columbia University's Earth Institute, says, "Fossil fuels get us coming and going. They damage ecosystems when we extract them from the earth. They cause damage to people and the environment when we burn them for energy."[33] Solar power is a much cleaner alternative to fossil fuels.

Solar Power Does Not Benefit the Environment

"Although solar is seen as clean energy in terms of carbon emissions, the production of many components is energy intensive and polluting."

—Jonathan Watts, the Asia environment correspondent for British newspaper the *Guardian*.

Jonathan Watts, "Solar Panel Factory Protests Tarnish China's Clean-Tech Efforts," *Guardian* (London), September 18, 2011. www.guardian.co.uk.

One of the main reasons solar power harms the environment is that it requires large areas of land. Although the sun does deliver vast amounts of energy to the earth, collecting that energy requires a large surface area. Solar panels can be placed on top of buildings, but utility-scale solar power plants would be needed to substantially increase the amount of solar power produced in the United States. These require large areas of land. Writer Bob Marshall emphasizes just how much land these plants take. He says, "Projects designed to produce power on scales large enough to light up communities do not resemble the photovoltaic panels on your neighbor's roof, nor that windmill on the local farm. They gobble up land on a stunning scale."[34] CSP plants, in particular, use far more land than power stations that burn fossil fuels. Robert Glennon, author and professor of law and public policy at the University of Arizona, reports that a 1,000 MW fossil fuel plant requires 640 acres (259ha), whereas a 1,000 MW CSP plant requires much more than that—about 6,000 acres (2,428ha).

Some people argue that comparisons such as this do not include the additional land required for mining fossil fuels such as coal. However, in the United States and some other places, much of the coal that is used in power plants comes from existing mines. This means that very little new land is being impacted by mining. According to Glennon, "Extant [existing] mines have the capacity to produce billions of tons of coal over the next century. Very few new coal mines are coming online."[35] Overall, says Glennon, solar power uses far more land.

Damaging the Natural Environment

In the vast tracts of land occupied by solar plants, the natural habitat is altered, and both people and animals are displaced. The Ivanpah solar power facility in California's Mojave Desert is an example of how a solar plant can do this. The Mojave Desert is ideal for a solar plant because it has miles of unused land and receives sunshine most days of the year. However, it is also home to the desert tortoise, a creature that has been in existence since the time of the dinosaurs, and the construction of the Ivanpah plant is killing tortoises and destroying their habitat. The project is located on 3,600 acres (1,457ha) of land, where thousands of mirrors focus the sun on solar receivers located on top of power towers. BrightSource Energy, the company building Ivanpah, has already reduced the size of the project and spent millions of dollars to protect and relocate the tortoises, but construction is still having a harmful impact on these creatures. The *Los Angeles Times* warns, "History has shown the tortoise to be a stubborn survivor, withstanding upheavals that caused the grand dinosaur extinction and ice ages that wiped out most living creatures. But unless current recovery efforts begin to gain traction, this threatened species could become collateral damage in the war against fossil fuels."[36]

In addition to covering large areas of land with mirrors or solar panels, many solar power plants also use large amounts of water for cleaning and cooling. The most efficient type of CSP plants are called wet-cooled plants, and according to Glennon, they use three to five times more water than coal or natural gas plants. This is of particular concern in the Southwest, where water is already scarce and droughts often occur. Competition for water pits wildlife against farmers, city dwellers, and recreational uses on a regular basis in the southwestern United States. Marshall points out just how environmentally harmful such water use can be to other sectors of society. He says, "Large scale solar farms will be located in areas where the sun shines most of the day—arid and desert regions of the West. Even a slight change in the water supply there can have disastrous effects on fish and wildlife."[37]

Manufacturing of Solar Components Causes Pollution

Solar power also harms the environment through the manufacturing of the various components needed to produce that power. Numerous toxic chemicals are used to make PV cells. These chemicals can leach into the environment during the manufacturing process and also when the PV components are disposed of at the end of their life. Once in the environment, these chemicals can contaminate food and water supplies, harming plants and both human and animal health.

The Silicon Valley Toxics Coalition, an organization that promotes environmental responsibility in technology, cautions that many new solar technologies use materials that have unknown environmental and health risks. It warns that increased production of solar panels could be adding large amounts of potentially toxic chemicals into the environment. The coalition says, "Little attention is currently being paid to the potential risks and consequences of scaling up solar PV cell production. The solar PV industry must address these issues immediately."[38]

> "Little attention is currently being paid to the potential risks and consequences of scaling up solar PV cell production. The solar PV industry must address these issues immediately." [38]
>
> —The Silicon Valley Toxics Coalition, an organization that promotes environmental responsibility in technology.

Lead is one harmful chemical associated with the solar power industry. Many solar systems rely on lead batteries to store the energy they create. Lead is a known environmental toxin; leakage often occurs when lead is mined and smelted and when batteries are manufactured and recycled. In a 2011 study published in the journal *Energy Policy*, researchers Perry Gottesfeld and Chris Cherry investigated 2 of the countries most involved in the mining of lead. They found that 33 percent of

Solar Power Uses Too Much Land

Solar power requires far too much land to generate a significant amount of power for it to be considered an environmentally friendly power source. Using so much land disturbs habitat and scenery. Other power sources require much less land for producing the same or more energy. In this chart, power plant size is given in megawatts, a common measure of power production. Both CSP and PV technology require many times more land than traditional power sources.

Plant Type	Plant Size (megawatts)	Land Area (acres)
Coal	500 – 1,000	640
Nuclear	500 – 1,000	640
Natural gas	200 – 500	320
	500	3,000
Concentrating solar	1,000	~6,000
Photovoltaic	1,000	12,160

Source: Robert Glennon, "Storm Clouds over Solar Energy?," *Solar Today*, April 2011. www.solartoday.org.

lead mined in China and 22 percent in India makes its way back into the environment. When lead leaks into the environment, it can cause numerous health problems, including learning problems in children and damage to the reproductive system, the cardiovascular system, and the central nervous system. Gottesfeld and Cherry predict that as China and India expand their solar power industries in coming years, there will also be a dramatic spike in lead pollution.

A 2011 incident in China illustrates how chemicals used in the manufacture of solar panels and other equipment can harm the environment. Solar panel factory JinkoSolar, in Zhejiang Province, was shut down for almost a month by the government after a four-day protest by several hundred villagers who charged that the company was releasing toxins into the river. According to the villagers, the toxins killed large numbers of fish in late August of that year and threatened their own health. Says one villager, Ren Suifen, "Since they set up their operations here, there has definitely been an impact on the villagers here. We do not know how our health will be in the future. This pollution is definitely harmful to us."[39] Government inspectors found contamination in the river, and JinkoSolar admitted that improper storage might have allowed pollutants to wash into the river from the factory during heavy rains.

Pollution from Used Components

In addition to pollution during the manufacturing process, solar components also have the potential to cause environmental pollution at the end of their life—a problem that is likely to surface in the near future. This is the finding of the Committee on Health, Environmental, and Other External Costs and Benefits of Energy Production and Consumption and the National Research Council, in an examination of the hidden costs of energy production and use. The two agencies say:

> Worn-out solar panels have potential to create large amount of waste, a concern exacerbated by the potential for toxic chemicals in solar panels to leach into soil and water. Many components of solar panels can be recycled, but the United States currently does not have or require a solar PV recycling system. . . . If solar energy for electricity were to become a significant part of the U.S. energy mix, more attention would need to be paid to damages resulting from the manufacture, recycling, and disposal of equipment.[40]

In fact, solar energy is becoming a more significant part of the energy mix both in the United States and around the world. As a result, solar panel manufacturers are creating millions of solar panels. When these panels become worn out in 20 to 30 years, there will be a huge amount of waste to dispose of, and the potential for a huge amount of pollution if that is not done properly.

Replacing One Harm with Another

Solar power development in the southwestern United States is a good illustration of the environmental concerns associated with solar power. A number of large-scale solar power installations that are currently under construction there have raised questions about whether the environmental impact will outweigh the benefits. For example, the Blythe Solar Power Project, near Blythe, California, will cover about 7,000 acres (2,833ha) and provide enough power for about 300,000 homes each year. The first phase of the project was to be completed in 2013. According to the project developer, Solar Trust of America, upon completion the project will be the largest solar power plant in the world. However, critics charge that the Blythe project and others like it will use huge amounts of land and water, destroying and displacing existing plants and animals. In addition, they assert that the environmental benefits will be minimal because solar is an inefficient form of power. Journalist Michael Haederle wonders, "Could . . . [these new solar installations] actually worsen the environmental catastrophe they are trying to avert?"[41]

> "If solar energy for electricity were to become a significant part of the U.S. energy mix, more attention would need to be paid to damages resulting from the manufacture, recycling, and disposal of equipment."[40]
>
> —The authors are part of the National Academy of Sciences, a nonprofit society of scholars dedicated to the furtherance of science and technology and to their use for the public good.

The general public, for the most part, sees solar energy as the ultimate in environmentally friendly, or green, technology. This public perception overlooks the reality of solar power's substantial environmental harms. In its rush to find alternatives to polluting energy sources such as coal, society may simply be trading one type of pollution for another.

Chapter Three

Can Solar Power Ever Replace Fossil Fuels?

Solar Power Can Replace Fossil Fuels

With fossil fuel supplies declining, the world needs an alternative power source. Every day, the earth receives enormous quantities of solar energy, and this is more than enough to replace fossil fuels. Solar power has the added advantage of not requiring connection to a power grid in order to bring energy to homes and businesses. Future cost reductions and technological improvements will make solar power a viable replacement for fossil fuels. And, unlike fossil fuels, solar power will never run out.

The Debate

Solar Power Is Not a Viable Replacement for Fossil Fuels

Despite the appeal of supplying all the world's energy needs from the sun, it is unrealistic and impractical to think that solar power could ever fulfill this need. As a life force solar is essential, but as a power source it is intermittent and therefore unreliable. Even if solar power can provide a portion of the world's electricity needs, it cannot replace fossil fuels in the crucial area of transportation. Solar-powered vehicles are simply not a realistic option for the future.

Solar Power Can Replace Fossil Fuels

"If harnessed properly, sunlight could easily exceed current and future electricity demand."

—James Hamilton, an economist in the Office of Occupational Statistics and Employment Projections at the US Bureau of Labor Statistics.

James Hamilton, "Careers in Solar Power," *Green Jobs: Solar Power*, US Bureau of Labor Statistics, June 2011, p. 1.

While the world currently burns fossil fuels to obtain the majority of its energy, this is not a viable long-term solution, because fossil fuels are a finite resource and will eventually be depleted. Eric McLamb is founder of the Ecology Global Network, an organization that works to educate people about preserving the environment. He explains why the world needs to look for an alternative to fossil fuels: "Fossil fuels formed from plants and animals that lived hundreds of millions of years ago and became buried way underneath the Earth's surface where their remains collectively transformed into the combustible materials we use for fuel." As a result, he says, "Fossil fuels are non-renewable. They are limited in supply and will one day be depleted. There is no escaping this conclusion."[42] The United States is one country that is particularly reliant on fossil fuels; it currently meets about 85 percent of its energy demand with them.

While critics contend that the world continues to discover new fossil fuel reserves, these new discoveries are not enough to meet the world's voracious demand for energy. Researchers James R. Fischer and Gale Buchanan state, "The known oil and gas fields are maturing, new fields are not being discovered fast enough to keep up with demand, and the new discoveries are in locations that are difficult to develop." For example, they report that the United States is using oil and gas at a far greater rate than new supplies are being discovered. According to the researchers, "The U.S. Geological Survey reported that the U.S. consumed 27 bb/y [billion barrels per year] in the last five years, but we only discovered 3 bb/y—a fraction of our consumption."[43] As fossil fuels become more scarce and expensive, society will need another abundant energy source to meet its energy needs.

A Vast Supply of Energy

Solar power is a good alternative to fossil fuels. The sun is a huge ball of hot gas that releases a tremendous amount of energy. A large amount of this energy makes its way to the earth, and this is more than enough to replace the energy that society gets from fossil fuels. According to the DOE, the earth gets enough energy from the sun every hour to meet its energy needs for a year. The DOE says, "Solar energy is by far the Earth's most available energy source. Solar power is capable of providing many times the total current energy demand."[44]

The sheer amount of energy available makes solar power the best option of all the possible replacements for fossil fuels. Researchers Richard Perez, Ken Zweibel, and Thomas E. Hoff compare world reserves of fossil fuels and the potential of various renewable resources and find that solar offers the most promise for meeting the world's energy demand. Because the solar resource is so vast, they argue, "Logic alone tells us, in view of available potentials, that the planetary energy future will be solar-based. Solar energy is the only ready-to-mass-deploy resource that is both large enough and acceptable enough to carry the planet for the long term."[45] Some parts of the world have greater solar resources than others; for example, those countries closest to the equator receive more energy from the sun. However, solar power has been successfully used all around the world. Germany, for example, is a world leader in solar power even though it receives far less sunshine than most parts of the United States.

> "Solar energy is by far the Earth's most available energy source. Solar power is capable of providing many times the total current energy demand."[44]
>
> —The DOE, a US government organization with the mission of advancing energy technology.

Cost Reductions Will Increase Viability

The future of solar power will include many innovations that will result in better technology and lower costs, making solar an even better replacement for fossil fuels. Solar power is still a relatively young technology compared to other power sources such as coal and natural gas. Given time to improve, it will become more efficient and less expensive, just as these other fuels have, and will be an increasingly attractive replacement for fossil fuels.

History already shows some dramatic improvements in solar and other new forms of renewable energy such as wind power. Letha Tawney, senior associate in the World Resources Institute's Climate and Energy Program, discusses this trend. She says that the solar PV modules manufactured today can create almost double the power for their size as those manufactured in 1982. Says Tawney, "Tremendous improvements have been made in how we site, operate, maintain, and integrate these power sources to squeeze every last drop of energy out of them."[46]

Steven Cohen, executive director of Columbia University's Earth Institute argues that society should not underestimate the power of technology to make dramatic and unimaginable improvements to solar power in the future. He points out that at one time, things such as iPods and cell phones were unimaginable, too, and argues that the future of solar power has similarly unimaginable innovations ahead. He says, "Over time, . . . [solar technology] will be subject to the same improvements and cost reductions that every consumer has seen in laptop computers and cell phones."[47] Such improvements will make solar a viable replacement for fossil fuels.

Power Without a Grid

In addition to being an abundant and powerful form of energy, solar power also has the advantage of providing power without a connection to a central power grid. Traditional forms of power such as coal-fueled electricity are created in a large power station, then transmitted to where they are needed through power lines. While some solar technology does involve building solar power stations and distributing power through a

Solar Power Has Greater Potential Energy than Fossil Fuels

Solar power has far more potential as an energy source than fossil fuels. This chart shows the total known reserves of natural gas, petroleum, and coal, as well as uranium, which is used to produce nuclear power. The volume of each sphere represents the total amount of energy estimated to exist. The largest sphere represents the total amount of energy that can be recovered from the sun in only one year. Just one year of solar power far exceeds all of the world's existing fossil and nuclear fuel reserves.

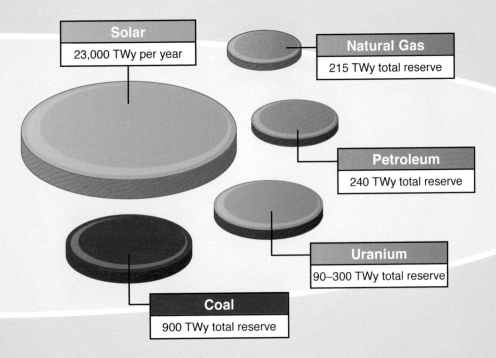

Solar

23,000 TWy per year

Natural Gas

215 TWy total reserve

Petroleum

240 TWy total reserve

Uranium

90–300 TWy total reserve

Coal

900 TWy total reserve

Note: TWy, or terawatt-years, is used to measure large quantities of energy.

Source: Richard Perez, Ken Zweibel, and Thomas E. Hoff, "Solar Power Generation in the US: Too Expensive, or a Bargain?," Atmospheric Sciences Research Center, August 11, 2011. www.asrc.cestm.albany.edu.

grid, a large percentage of solar technology involves creating power without connection to a grid. A house or a business can install solar panels on its roof and obtain power directly from those panels rather than from a power grid. Or a solar water heater can be operated from solar power rather than electrical or gas power from the grid. This eliminates the need for expensive power transmission lines.

David Crane, CEO of NRG Energy, one of the largest electric companies in the United States, believes that this type of solar power—distributed solar power—will explode in the near future because there is so much potential and room for growth. He says, "We believe that in the next 3 to 5 years you'll be able to get power cheaper from the roof of your house than from the grid. Solar is going to go from this thing that right now is like .1 percent of the market to 20 to 30 percent of the overall electricity mix. That's huge."[48] Haiti is one country that is utilizing distributed solar power. According to the country's president, Michel Martelly, only 30 percent of the country's population currently has access to electricity. One new program that is part of efforts to change this involves giving rural Haitians loans for small solar kits that will give them electricity to charge things such as cell phones and computers. Within two years, Martelly hopes, this program will double the number of rural households that have access to electricity.

> "Solar is going to go from this thing that right now is like .1 percent of the market to 20 to 30 percent of the overall electricity mix. That's huge."[48]
>
> —David Crane, CEO of NRG Energy, one of the largest electric companies in the United States.

In addition to providing power for individual homes and businesses, distributed solar power has the potential to be used for numerous situations where a small amount of power is needed and there is no connection to a power grid. Researchers Fischer and Buchanan give some examples: "A unique and valuable aspect of photovoltaic solar power is its application to small-scale and remote situations, such as opening and closing gates, illuminating signs, powering electric fences, pumping water, and providing ornamental illumination for gardens and walkways."[49] *USA Today* reports that an increasing number of towns and cities are using solar-powered school safety signs and road warning signs as a way to save both power and money.

A Viable Alternative

The South Pacific islands of Tokelau are already showing that it is viable to use solar power to replace fossil fuels. In the past, these 3 small islands, with about 1,500 citizens, have used imported fossil fuels for energy. However, Tokelau announced that it aimed to produce 93 percent of its electricity with solar power by 2012. Batteries were to store excess electricity for use at night. During periods of cloud cover or when demand exceeds supply, generators running on coconut oil were to provide additional power. Foua Toloa, the head of Tokelau, hoped the example of this island nation would show the potential of solar power and inspire other countries to undertake similar projects. He says, "We stand to lose the most of any country in the world due to climate change and the rising sea levels, so leading the way . . . is a message to the world to do something."[50]

Solar power is already a viable replacement for fossil fuels. The sun is a tremendous source of energy that is already used successfully for power around the world. In the words of the Solar Energy Industries Association, "No scientific breakthroughs are required for solar energy to power America. Solar is ready and available today."[51]

Solar Power Is Not a Viable Replacement for Fossil Fuels

"It would be cost prohibitive to make solar energy mainstream for major world consumption in the near future."

—Eric McLamb, founder of the Ecology Global Network, an organization that works to educate people about preserving the environment.

Eric McLamb, "Fossils Fuels vs. Renewable Energy Resources," Ecology Global Network, September 6, 2011. www.ecology.com.

Fossil fuels provide a steady, reliable source of energy that is available day and night, through all seasons, and in most all inhabited corners of the earth. When demand for electricity is high, fossil-fueled power generators increase output. When demand falls, they decrease output. Solar power does not work this way; it cannot meet the world's energy needs now, and it will not meet this need in the future. Solar power is not a viable replacement for fossil fuels.

One of the weaknesses of solar power is that levels of production depend on the availability of sunlight. At best, a solar power station might get sunshine all day and thus produce power all day long. It will not be able to produce when the sun is not shining at night or when clouds cover the sun. And it is not a good option in places that get little sunshine during the day. The reality is that the output of solar power fluctuates according to the time of year, time of day, and weather conditions. In contrast, fossil-fueled power can be produced at any time, regardless of weather conditions or time of day. David Bergeron, president of SunDanzer Development, argues that because of this inconsistency, "The bottom line is that solar cannot be counted on when power is needed."[52]

Solar Power Requires Too Much Land

As well as being intermittent and unreliable, solar power is also inferior to fossil fuels because it requires more land area to produce the same amount of energy that fossil fuels do. This is a concept known as density. Fossil fuels are defined as a very dense form of energy because they require a relatively small area in order to produce a large amount of electricity. In contrast, solar power and other renewable forms of energy like wind power require a large surface area to collect energy. Benjamin Zycher, author of *Renewable Electricity Generation: Economic Analysis and Outlook*, explains why. He says:

> The energy content of wind flows and sunlight . . . is far less concentrated than that of the energy contained in fossil or nuclear fuels. To compensate for this unconcentrated nature of renewable energy sources, the facility operator or power utility must invest large amounts in land and materials to make renewable generation technically practical for generating nontrivial amounts of electricity.[53]

In order to produce the same amount of energy as a fossil fuel power station that takes up a few hundred acres, a solar power station must cover thousands of acres of land with panels or mirrors. So replacing fossil fuels with solar power means that a greater area of land must be devoted to power production in order to produce the same amount of power.

In a 2011 article, Robert Glennon, an author and professor of law and public policy at the University of Arizona, compares the land requirements of fossil fuel and solar power plants. He finds that a solar power plant requires more than 6,000 acres (2,428ha) to produce the same amount of power that a fossil fuel plant can in less than 1,000 acres (405ha). In addition, he adds, most solar power plants produce substantially less power than fossil fuel plants, due to intermittency problems. This means that even more land is required to generate the same amount of power as a fossil fuel plant. Glennon says, "If we focus on the megawatt-hours actually produced, we would have to multiply the land footprint of solar plants by three to five times in order for solar plants to generate the same number of megawatt-hours as traditional fossil fuel plants."[54] Society cannot afford to devote so much land to producing power, since that land is also needed for many other things such as growing food crops.

Will Not Satisfy World Energy Needs

Because it is less efficient than fossil fuels, solar power cannot provide enough power to meet current demand, much less keep pace with the world's rapidly growing energy needs. The world's population is increasing every year and will require increased amounts of energy to support it. In addition, as many countries become more developed, their citizens are consuming increasing amounts of energy. For example, as India has developed in recent years, its middle class has grown, and as a result there has been an increase in the number of people who can afford items such as washing machines, refrigerators, and air conditioners. Such items also mean increasing energy demands. To meet these increasing demands, society needs more efficient ways of producing energy, not less efficient ones. The EIA predicts that world electricity demand will more than double between 2007 and 2050. Patrick J. Michaels, senior fellow in environmental studies at the Cato Institute, argues that society should be looking for more dense sources of energy, not a less dense one. He says, "Both solar and wind stand in the way of the torrent of history. The global trend towards increasingly dense energy . . . is obvious and logical. As economies develop they consume enormous amounts of energy."[55]

> "Both solar and wind stand in the way of the torrent of history. The global trend towards increasingly dense energy . . . is obvious and logical. As economies develop they consume enormous amounts of energy."[55]
>
> —Patrick J. Michaels, a senior fellow in environmental studies at the Cato Institute, a public policy research organization.

China is an example of how difficult it will be to meet the world's rising energy demand with solar power. In recent years China has greatly

Solar Power Does Not Produce Enough Electricity to Replace Fossil Fuels

A solar thermal power plant only produces a fraction of the energy produced by a coal or natural gas plant, making it a poor replacement for fossil fuels. This chart compares the average yearly output for solar thermal, coal, and natural gas power plants. Output is measured in million megawatt-hours of electricity produced, a common way to measure electrical power. The chart reveals that coal and natural gas plants are far more efficient, producing more than four times as much power as solar thermal.

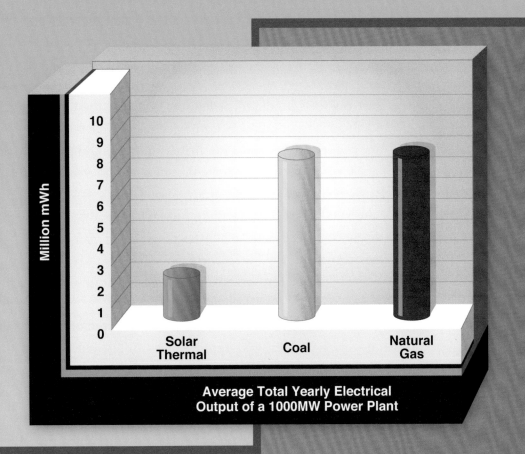

Source: Institute for Energy Research, "Salazar Applauds—but Solar Project Will Be a Fraction of Its Competitors," June 22, 2011. www.instituteforenergyresearch.org.

increased its use of renewable energy, including solar. Yet even with these dramatic increases, not only does China still need fossil fuels, but it will also need to increase the quantity of fossil fuels it uses. Journalist Ronald Brownstein says, "China's energy demands are growing so fast that despite this ambitious clean-energy push, the country is still projected to double its consumption of coal by 2030."[56]

Not Compatible with the Current Power Grid

Another important issue that will prevent solar power from replacing fossil fuels is that the existing power grid was not designed for collecting and distributing large amounts of solar power. If solar power plants are to provide a significant amount of energy, improvement and expansion of the existing power grid, such as building new power lines to the places where solar power is produced, would be required. The Center for Climate and Energy Solutions points out, "While estimated solar resources are vast, frequently the areas with the most ideal conditions for utility-scale solar electricity generation are remote and far removed from end-users of electricity. In particular, the U.S. Southwest possesses enormous solar resources but lacks transmission to transmit large amounts of solar power."[57] Even where transmission networks do exist, in many cases they are outdated and poorly suited to solar power. Making the necessary improvements would cost millions of dollars and make solar power too expensive to be viable.

Even using large amounts of distributed solar power instead of power stations will require significant grid improvements. While individual homes and businesses using solar do not necessarily need to be connected to the grid, in most cases they are connected in order to obtain energy when their solar system is not providing enough, or to sell energy to the power company when they are producing extra. The *Deloitte Review* argues that such a connection will require grid improvements. It says, "Distributed solar will require grid operators to install technology to monitor power supply and demand, balancing thousands of individual generators with central power plants."[58]

Transportation

One of the biggest problems with trying to replace fossil fuels with solar power is in the area of transportation. The world uses large quantities of oil to fuel cars, trucks, planes, and other types of vehicles, and solar power is not a viable replacement for that oil. According to the EIA, in the United States more than 95 percent of all the energy used for transportation comes from oil. Solar power vehicles do exist; however, they are not cost-effective or efficient enough for widespread use. The *Solar Power Blog* says, "Solar cars are in existence now, but don't expect to find one at a car lot any time soon. These cars are mostly made by scientists for competitions or for research purposes. As it stands now, solar cars are not ready for street use. This is mostly because they can't harness enough energy from the sun to really run without problems."[59]

> "Ultimately . . . using sunlight to produce electricity will never supply enough of the energy we need."[60]
>
> —David Rotman, a science and business journalist and editor of the magazine *Technology Review*.

Designing a practical and affordable solar vehicle remains an obstacle for the future. Current solar car designs are expensive. What's more, because they must be constructed to be aerodynamic and efficient they have little room for passengers. This makes them an impractical alternative to traditional cars. Overall, society does not yet have the technology to replace oil-fueled cars with solar-powered ones.

Society Needs a Steady Power Source

Solar power cannot replace fossil fuels as a major provider of the world's energy needs. Society needs a steady power source that provides power on demand and can match its rapidly growing energy demands. As journalist David Rotman argues, there are just too many problems with solar power to make it a viable replacement for fossil fuels. He says, "Ultimately . . . using sunlight to produce electricity will never supply enough of the energy we need."[60]

Chapter Four

Should Government Play a Role in Developing Solar Power?

Government Should Help Develop Solar Power

Throughout history governments have helped develop numerous energy sources, and solar power should be no exception. Government involvement spurs private investment and accelerates development. It also helps the United States remain competitive in the world solar power market. Germany's successful solar industry proves the value of government support for solar power. This technology will be beneficial to society, and the US government should use its power to make sure that it succeeds.

The Debate

Government Should Not Be Involved in Developing Solar Power

If solar power is a viable technology that makes financial sense for consumers, it will naturally succeed in the free market. However, when the government becomes involved in that market, market forces are distorted, and solar technology might succeed through political influence and government funding even when the technology is not good. Excessive government financial help has deterred innovation in solar technology by allowing solar developers to succeed without much risk. It has also led to increased energy prices for consumers.

Government Should Help Develop Solar Power

"Government should accelerate its efforts and do more to develop and support clean energy."

—Vidya Kale, an optoelectronic engineer who lives in Oregon.

Vidya Kale, "Government Support for Renewable Energy Development Must Continue," OregonLive.com, September 8, 2011. www.oregonlive.com.

In the United States and many other countries, the government is involved in the development of solar power in a variety of ways. Government involvement takes the form of grants, subsidies, and loan guarantees that all make it cheaper for companies to invest in solar power. The government also helps solar power through tax policy. For example, in the United States the federal Investment Tax Credit, in place until 2016, gives individuals or businesses that invest in solar power a tax credit equal to 30 percent of the cost of the solar system. Yet another way the government helps solar power is through renewable portfolio standards, which are requirements that solar or other renewable energy sources constitute a specific percentage of a utility's total mix of energy sources. A number of US states have such requirements. For example, California law requires that by 2020 a third of its power comes from renewable energy. Governments also help solar power through feed-in tariffs, which are long-term contracts for producers of solar power and other types of energy and are based on the price of producing that energy. Feed-in tariffs spur investment in solar power because they guarantee solar power producers a reasonable return on their investment for the long term. Germany and Spain have used feed-in tariffs to significantly increase their production of solar power. Overall, the United States and other governments invest millions of dollars in solar power every year.

Government Investment Spurs Private Investment

The availability of government grants, loans, and other financial incentives makes it easier and less risky to enter the solar power business. Thus, by investing in solar power, the government is encouraging private industry to do the same. To illustrate this, the Solar Energy Industries Association gives the example of the federal Loan Guarantee Program, which provides loan guarantees for innovative energy technologies. The association maintains that by providing loans to solar power and other alternative energy technologies, the program spurs millions of dollars of private investment in these technologies, too. "The Department of Energy Loan Guarantee Program helps make solar power cheaper and more affordable for businesses and homeowners," says the association. "Each program dollar leverages $13 in private investment. As of September 16, 2011, DOE has made commitments to 42 energy projects, sparking private investment of more than $40 billion."[61]

In his 2012 State of the Union speech, President Barack Obama argued that government investment in renewable energy technologies such as solar power has been very successful and will continue to be so. He said, "Because of federal investments, renewable energy has nearly doubled, and thousands of Americans have jobs because of it." He concluded, "It's time to . . . double-down on a clean energy industry that has never been more promising."[62]

> "Because of federal investments, renewable energy has nearly doubled, and thousands of Americans have jobs because of it."[62]
>
> —Barack Obama, forty-fourth president of the United States.

Waiting to invest in solar technology until it has matured is not an option, because government involvement is one of the major factors driving solar's development. Government policy and financial help are spurring the development that will help solar power become a mature technology. As Harry Gray of Caltech Solar Power argues, the government must take the lead and set an example for the rest of society. He says, "We can't sit back and wait for breakthroughs. We need to show people that solar can work."[63]

Keeping Up with the Rest of the World

Government investment in solar power also helps ensure that the United States does not fall behind other countries in developing this technology, a failure that could be economically harmful. Some countries give solar power significant government support, which allows the industry to grow and take a lead in the world market. For example, in recent years the Chinese government has given the solar industry large subsidies. As a result, Chinese manufacturers can produce solar panels more cheaply than many other countries, and China has become a leading manufacturer of solar panels. Rick Peters is an engineer at Solar Energy Services in Millersville, Maryland, and says that Chinese solar panels are about 10 percent cheaper than US ones. Because most customers want the lowest-price panels available, Peters says, "probably about 70 percent of what we install is Chinese panels."[64] This means that China is receiving the economic benefits of manufacturing solar panels, such as jobs and income, while the United States is missing out.

Gordon Brinser, the president of SolarWorld, warns that the United States needs to take action to regain those economic benefits. He says, "We need to wake up as a nation because we're on the verge of losing one of the core industries I believe for our future. Right now, we're dependent on Middle East for our fossil fuel oil. And very shortly, we're going to be dependent upon on the [Far East] for our solar."[65]

Obama also argues that if the United States does not support its solar power industry, it will lose it. He says, "Even if the technology was developed here in the United States, they end up going to China because the Chinese government is saying, 'We're going to help you get started, we'll help you scale up, we'll give you low-interest loans or no-interest loans. . . . We will do whatever it takes for you to get started here.'"[66]

A Job for Government

Energy is an important part of a thriving society, and this is another reason that the government should take a role in developing affordable and reliable energy sources like solar power. Developing these sources will benefit society. The US government has a history of helping energy

A Large Number of Americans Believe the Government Should Support Solar Power

This survey, conducted by manufacturer SCHOTT Solar, reveals that a large number of Americans favor government support for solar power. Of one thousand American adults surveyed 82 percent said they support federal tax credits and grants for the solar industry. When respondents were asked to choose an energy source they would support if they were in charge of US energy policy, 39 percent chose solar, making it the top choice. Only 16 percent said they do not believe the government should invest in energy sources.

If you were in charge of US energy policy and could choose to provide financial support for one of the following energy sources during your term in office, which would you choose?

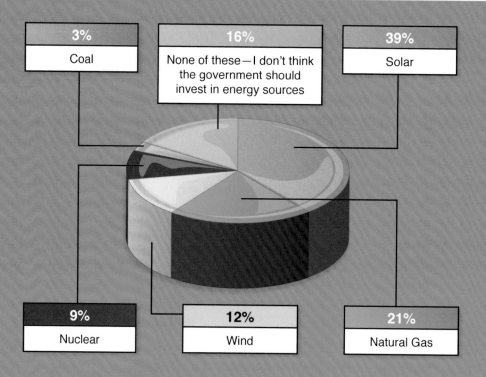

3%	16%	39%
Coal	None of these—I don't think the government should invest in energy sources	Solar

9%	12%	21%
Nuclear	Wind	Natural Gas

Source: SCHOTT Solar, "4th Annual SCHOTT Solar Barometer Shows 9 Out of 10 Americans Support Solar," November 1, 2011. www.schottsolar.com.

technologies develop. Richard W. Caperton, director of Clean Energy Investment at the Center for American Progress, explains how the government has played a central role in US energy development:

> The federal government has made investments in energy for more than a century, by granting access to resources on public lands, helping build railroads and waterways to transport fuels, building dams to provide electricity, subsidizing exploration and extraction of fossil fuels, providing financing to electrify rural America, taking on risk in nuclear power, and conducting research and development in virtually all energy sources.[67]

The United States, it could be argued, has widespread access to affordable and reliable electricity precisely because of government involvement in the development of all of these different industries. Caperton concludes that continued government involvement will be important if the country is to meet its future energy needs. He says, "It's clear that the smart choice is to make these investments to meet the next generation of energy challenges and to produce a foundation of affordable, reliable, and clean energy alternatives for future waves of investment and opportunity."[68]

"The federal government has made investments in energy for more than a century."[67]

—Richard W. Caperton, director of Clean Energy Investment at the Center for American Progress, a public policy research and advocacy organization.

In addition to providing an affordable and reliable source of energy for the future, solar power benefits society in other ways, too. Current sources of energy such as coal and oil pollute the environment and harm human and animal health. Replacing these with nonpolluting solar power can reduce both the monetary and human costs caused by such pollution. Solar power can also enhance energy security because it can be domestically produced and thus reduce the need to rely on imported fossil fuels.

The German Example

Germany's experience with solar power shows the value of government involvement. The German government has been heavily involved in encouraging the growth of solar power there. It has done this through feed-in tariffs, where producers of solar power electricity are guaranteed to receive fixed rates for the electricity they produce for 20 years from the date of installation. These feed-in tariffs started in 1991, well before they were used in other countries. The result of the tariffs has been a small increase in the price of power but record growth in solar power installation in that country. According to BSW-Solar, Germany's main solar industry association, solar power will make up 7 percent of the country's energy mix by 2016, up from 4 percent in 2012.

Germany's actions have also been beneficial to solar development around the world. By using so much solar power, it has created an increased demand for solar panels, and as a result manufacturers have been able to mass-produce the panels. Mass production results in a cheaper price per panel, so the result has been a reduction in the overall price of solar worldwide. Journalist Evan I. Schwartz argues that Germany proves the success of government involvement in encouraging solar power development. He says, "The German experiment does show that a large industrial society can reach ambitious goals for scaling up new sources of clean electricity." As a result of the government encouraging the development of solar power, says Schwartz, "Germany expects to produce most of its electricity from renewable sources [including solar power] by 2030."[69]

Government involvement is vital to the success of any new energy industry. This has been true in years past and will continue to hold true for the future. As the Solar Energy Industries Association explains, "Every major energy source and technology has benefited from federal government R&D [research and development] support and incentives of various types."[70] Solar power should be no exception. With government support, it can become an important part of the world's energy supply in the future.

Government Should Not Be Involved in Developing Solar Power

"We are an energy rich nation and a nation that leads the world in innovation. Private capital, market forces, our abundance of energy and economic and energy realities can lead to the development of new energy sources in time. Trying to force them into the market [through government action] won't work."

—William O'Keefe, CEO of the George C. Marshall Institute, a think tank that focuses on scientific issues and public policy.

William O'Keefe, "Let Markets Shape Our Energy Future," September 26, 2011, response to Amy Harder, "What Role Should Government Play in Energy Production?," *Energy Experts Blog, National Journal*, September 26, 2011. www.nationaljournal.com.

Government involvement in solar power distorts free-market forces. These forces, not government policy, are the best way to determine whether a particular technology should be in existence. In the free market, a technology can succeed if it is innovative and affordable; otherwise, it may fail. This should be how the future of solar power is decided. If it proves to be innovative and affordable, it will succeed in the free market. If the government becomes involved, offering financial and policy help to ensure the success of this technology, it is ignoring the question of whether solar power is actually viable. Nicolas Loris, a policy analyst at the Heritage Foundation, argues that the question of viability should be left to the market, not put in the hands of the government. He says, "No evidence exists to suggest that the government has better knowledge to make investment decisions or to commercialize technologies when the private sector chooses to bypass these opportunities. If there is a role for alternative sources in America's energy portfolio, it should be dictated by price and competition, not government handouts."[71]

Government Involvement Leads to Higher Electricity Rates

Government financial help for solar power means that consumers ultimately pay more in taxes to finance that help, and in utility bills since solar power is more costly than fossil fuel power. This chart compares average retail electricity prices (in euros) in some European countries. It reveals that Denmark, Germany, Italy, and Spain, countries that have given substantial subsidies to solar and other renewable types of energy in recent years, have much higher electricity prices than some other European countries.

European Country	Average User (euro cents per kWh)
Denmark	35.699
Germany	33.089
Italy	28.217
Austria	28.087
Belgium	27.492
Luxembourg	26.854
Netherlands	25.158
Spain	24.290
Ireland	24.853
Hungary	24.606
Portugal	24.186
Cyprus	24.157
Slovakia	22.707
Malta	21.330
Czech Republic	21.098
Sweden	21.025
Poland	19.677
UK	19.532
Finland	18.662
France	17.618
Latvia	15.791
Greece	15.385
Romania	14.892
Lithuania	13.819
Estonia	13.456
Bulgaria	12.543

Source: Elliot H. Gue, "Popping the Green Bubble," *Investing Daily*, November 1, 2011. www.investingdaily.com.

David Milroy, CEO of Vertichem, a company that produces environmentally friendly chemicals, points out that while government support might be helpful, it is not what will make new renewable energy technology succeed. Instead, he says, "It is a process that requires investment and strategic placement of private monies—not only government subsidies—and the business acumen and leadership skills to present it to the market in a compelling and attractive way."[72] Milroy maintains that innovation and investment from the free market are the key to making green technologies such as renewable energy succeed.

Government Involvement Means Political Influence

When the government becomes involved, the future of a technology such as solar power is influenced by other things, such as political considerations, rather than the quality of the technology. William O'Keefe, CEO of the George C. Marshall Institute, explains the way that politics influences federal investment: "Subsidies—whether loan guarantees or direct handouts—take taxpayer dollars and give them to companies that are doing what the government wants instead of what the market wants."[73] According to O'Keefe, where these subsidies go often depends on which companies are best able to persuade the government to subsidize them, and not so much on the product they are creating. The result, he says, is "a class of businesses that profit from skill in the regulatory and lobbying arenas instead of the marketplace."[74]

In the United States, government financial help for solar power has resulted in a solar power industry that is not economically viable and does not even contribute a significant amount of electricity. In recent years the government has offered large tax breaks and subsidies for solar power. Additionally, many states have passed legislation requiring utilities to buy a significant part of their power from renewable

> "Subsidies—whether loan guarantees or direct handouts—take taxpayer dollars and give them to companies that are doing what the government wants instead of what the market wants."[73]
>
> —William O'Keefe, CEO of the George C. Marshall Institute, a think tank that focuses on scientific issues and public policy.

sources such as solar. As a result, companies are guaranteed a profit regardless of the type of product they produce, because they receive government help to produce it and have a guaranteed market to sell to. For example, in California, utilities must buy 33 percent of their power from clean-energy sources by 2020. Kevin Smith, chief executive of solar company SolarReserve, says, "It's like building a hotel, where you know in advance you are going to have 100 percent room occupancy for 25 years."[75]

Because companies do not necessarily need to produce a good product to make a profit, in many cases they have not. Instead, there has been a proliferation of solar power companies that are not economically viable and do not produce a significant amount of power. Overall, argues the British newsweekly *The Economist*, the solar industry has received far too much government help. *The Economist* maintains, "The rush to subsidise solar power over the past decade has been massively wasteful."[76] In some cases the easy availability of subsidies deters innovation because companies are in a rush to go into business now with existing technology so they can take advantage of the subsidies, rather than spending time on research and development of an improved product first.

A Risk-Free Investment

In a 2011 report, the *New York Times* discusses the example of the California Valley Solar Ranch project, located in San Luis Obispo, California. The project, which began construction in late 2011, is a PV solar system that will provide enough energy to power 100,000 homes. The cost of the solar ranch is estimated at $1.6 billion, according to the *New York Times*. If the project fails, however, NRG Energy (the company constructing it) does not risk losing $1.6 billion of its own money, because much of the project will be funded by the government. The federal government guaranteed a low-interest loan for construction, and when the project is complete, NRG will also be eligible to receive 30 percent of its cost back as a cash grant. In addition, under state law the company will not have to pay property taxes to the county. And finally, because of state mandates that California utilities must buy 33 percent of their power from clean sources by 2020, the project has a 25-year contract with the local utility. So due to government funds and mandates, the developers of the solar ranch face less risk of losing money and have a guaranteed future market for their power, regardless of whether or not they can produce it at a reasonable price.

Increased Energy Prices

Another way that government involvement harms consumers is by forcing them to pay increased prices for power. The California Valley Solar Ranch project illustrates this. Local utilities there must soon buy 33 percent of their power from clean sources such as the solar ranch. Solar power is a more expensive form of power, so when the government requires increased use of solar power, such as that from California Valley Solar Ranch, it costs consumers more in both taxes and utility bills. According to *Sacramento Bee* columnist Dan Walters, as a result of government mandates in California, customers will pay increasingly high rates for power. He says, "California's average retail electric rate of 13.24 cents per kilowatt-hour is already the ninth highest in the nation, 50 percent above average. And when those 'renewable portfolios' [where utilities buy 33 percent of their power from renewable sources] come online, power bills will ratchet rapidly upward."[77]

In other places where the government has offered financial incentives for the increased use of solar power and other renewables, consumers frequently pay higher energy bills. In 2010 the United Kingdom implemented a feed-in tariff for solar power, meaning that solar power producers get long-term power contracts and are paid higher prices for their power, since it is more expensive to produce than power from other sources such as coal. The result, says Patrick J. Michaels, senior fellow in environmental studies at the Cato Institute, is that "electricity prices have gone through the roof."[78]

> "We've spent billions on technology and research and subsidies, and it's still the most expensive way of generating electricity."[82]
>
> —Tom McClintock, a Republican representative for California.

In a 2012 interview, German politician Michael Fuchs argued that in Germany, government subsidies for solar power have cost consumers too much. While subsidies have helped the solar industry grow there, they have been expensive for consumers, he said. In fact, Fuchs insisted, "We have to limit expansion in solar electricity. . . . Otherwise electricity will become unaffordable."[79]

Solyndra

The story of California solar power company Solyndra illustrates the harms of government involvement in the industry. Solyndra manufactured cylindrical solar panels—a new type of technology for the solar industry—and received $527 million in federal loans. In 2010 when Obama visited the Solyndra factory in Fremont, California, he claimed that this loan and other government help for renewable energy would offer significant economic benefits. Obama said at the time, "When it's completed in a few months, Solyndra expects to hire a thousand workers to manufacture solar panels and sell them across America and around the world. And this in turn will generate business for companies throughout our country who will create jobs supplying this factory with parts and materials."[80]

Yet the next year, unable to compete with flat panels made in China, Solyndra declared bankruptcy. Attorney Jonathan W. Emord maintains that Solyndra's failure reveals the harms of government involvement in solar power. He says, "The political decision to bet tax dollars on Solyndra reveals the inherent inability of politicians to outsmart the market in deciding the U.S.'s future." Government decisions are often based on politics, not market expertise, says Emord, and for that reason they frequently fail. He says, "The arrogant assumption in Washington is that politicians know better than we do how to best spend our money."[81] Emord maintains that rather than government taking taxes from the US population and using them to subsidize unsound companies such as Solyndra, the money should be left in the private sector, where private investors with more expertise than the government will make better business decisions about how to use it.

Viable Technologies Will Succeed in a Free Market

Overall, government involvement in solar power has not been beneficial and should not continue. Says Representative Tom McClintock, "We've spent billions on technology and research and subsidies, and it's still the most expensive way of generating electricity."[82] Solar power should be left to the forces of the free market, and if it is a viable technology for society, that will be proved without the distortions of political influence.

Source Notes

Overview: Visions of the Future: Solar Power

1. BSW-Solar, "Statistic Data on the German Solar Power (Photovoltaic) Industry," October 2011. www.solarwirtschaft.de.
2. Union of Concerned Scientists, "How Solar Energy Works," December 16, 2009. www.ucsusa.org.
3. Solar Energy Industries Association, "Industry Data." www.seia.org.
4. European Photovoltaic Industry Association, *Market Report 2011*, January 2012. www.epia.org.
5. Renewable Energy Policy Network for the 21st Century, "Renewables 2011: Global Status Report," 2011. www.ren21.net.
6. Olaf Babinet et al., "Solar's Push to Reach the Mainstream," *Deloitte Review*, January 18, 2011. www.deloitte.com.

Chapter One: Is Solar Power Affordable?

7. Paul Krugman, "Here Comes the Sun," *New York Times*, November 6, 2011. www.nytimes.com.
8. Steven Cohen, "Solar Power and the Future of Fossil Fuels," *Huffington Post*, June 20, 2011. www.huffingtonpost.com.
9. Elizabeth Svoboda, "Debunking the Top 10 Energy Myths," *Popular Mechanics*, July 7, 2010. www.popularmechanics.com.
10. Svoboda, "Debunking the Top 10 Energy Myths."
11. Quoted in ABC15.com, "Is Solar Technology Just Too Expensive?," November 7, 2010. www.abc15.com.
12. Quoted in *Yale Environment 360*, "A Power Company President Ties His Future to Green Energy," November 9, 2011. http://e360.yale.edu.
13. Andrea Luecke, "Data Shows Solar Investment Creates Jobs," October 1, 2011, response to Amy Harder, "What Role Should Government Play in Energy Production?," *Energy Experts Blog, National Journal*, September 26, 2011. www.nationaljournal.com.

14. Quoted in Alan Harman, "Running on Sunshine: Solar Power Runs This Chicken Farm," *Countryside & Small Stock Journal*, November/December 2011, p. 62.

15. Richard Perez, Ken Zweibel, and Thomas E. Hoff, "Solar Power Generation in the US: Too Expensive, or a Bargain?," Atmospheric Sciences Research Center, August 11, 2011. www.asrc.cestm.albany.edu.

16. Quoted in ABC15.com, "Is Solar Technology Just Too Expensive?"

17. Babinet et al., "Solar's Push to Reach the Mainstream."

18. Quoted in ABC15.com, "Is Solar Technology Just Too Expensive?"

19. Quoted in Ed Crooks, "Sun Sets on BP's Solar Hopes," *Financial Times*, May 13, 2009. www.ft.com.

20. BP, "Solar Power." www.bp.com.

21. Quoted in ABC15.com, "Is Solar Technology Just Too Expensive?"

22. Matthew L. Wald, "Solar Power Industry Falls Short of Hopes in Job Creation," *New York Times*, October 25, 2011. www.nytimes.com.

23. Vance Ginn, "Solar Energy: An Expensive Path to Job Creation," *PolicyMic*, 2011. www.policymic.com.

24. Center for Climate and Energy Solutions, "Solar Power," August 2011. www.c2es.org.

Chapter Two: How Does Solar Power Impact the Environment?

25. US Energy Information Administration, "Nonrenewable Coal." www.eia.gov.

26. Union of Concerned Scientists, "How Natural Gas Works," August 31, 2010. www.ucsusa.org.

27. Elizabeth Shogren, "Air Quality Concerns Threaten Natural Gas's Image," NPR, June 21, 2011. www.npr.org.

28. Monique Hanis, "To Fight Big Pollution, We Need Big Solar," *Earth Island Journal*, Autumn 2010, p. 56.

29. Mona Reese, "The Environmental Cost of Solar Panel Manufacturing," *Renewable Energy World.com*, February 12, 2011. www.renewableenergyworld.com.

30. Hanis, "To Fight Big Pollution, We Need Big Solar."

31. Hanis, "To Fight Big Pollution, We Need Big Solar."

32. Solar Trust of America, "Habitat Conservation." http://solartrustofamerica.com.

33. Cohen, "Solar Power and the Future of Fossil Fuels."

34. Bob Marshall, "Green Energy Land Rush: Large-Scale Wind and Solar Projects Could Exact a Toll on Wildlife," *Field & Stream*, December 2009. www.fieldandstream.com.

35. Robert Glennon, "Storm Clouds over Solar Energy?," *Solar Today*, April 2011. www.solartoday.org.

36. Julie Cart, "Saving Desert Tortoises Is a Costly Hurdle for Solar Projects," *Los Angeles Times*, March 4, 2012. www.latimes.com.

37. Marshall, "Green Energy Land Rush."

38. Silicon Valley Toxics Coalition, "Toward a Just and Sustainable Solar Energy Industry," January 14, 2009. http://svtc.org.

39. Quoted in Royston Chan, "China Firm Apologizes for Toxic Waste After Protest," Reuters, September 20, 2011. www.reuters.com.

40. Committee on Health, Environmental, and Other External Costs and Benefits of Energy Production and Consumption and the National Research Council, *Hidden Costs of Energy: Unpriced Consequences of Energy Production and Use.* Washington, DC: National Academies, 2010, pp. 144–45.

41. Michael Haederle, "Solar Showdown: Are New Solar Power Projects Anti-environmental?," *Miller-McCune*, April 18, 2011. www.miller-mccune.com.

Chapter Three: Can Solar Power Ever Replace Fossil Fuels?

42. Eric McLamb, "Fossils Fuels vs. Renewable Energy Resources," Ecology Global Network, September 6, 2011. www.ecology.com.

43. James R. Fischer and Gale Buchanan, "Developing Solar Power," *Resource*, May/June 2010, p. 15.

44. US Energy Information Administration, "Solar Explained," June 6, 2011. www.eia.gov.

45. Perez et al., "Solar Power Generation in the US."

46. Letha Tawney, "Want Low-Cost Clean Energy? Bank on Innovation," *WRI Insights*, November 14, 2011. http://insights.wri.org.

47. Cohen, "Solar Power and the Future of Fossil Fuels."

48. Quoted in *Yale Environment 360*, "A Power Company President Ties His Future to Green Energy."

49. Fischer and Buchanan, "Developing Solar Power."

50. Quoted in R.E. Christian, "Renewable Energy to Power Tiny Island Completely," *Online Journal*, December 14, 2011. http://onlinejournal.com.

51. Solar Energy Industries Association, "Solar Energy Facts: Correcting Old Myths," September 23, 2011. www.seia.org.

52. Quoted in ABC15.com, "Is Solar Technology Just Too Expensive?"

53. Benjamin Zycher, "Wind and Solar Power, Part I: Uncooperative Reality," American Enterprise Institute, January 17, 2012. www.aei.org.

54. Glennon, "Storm Clouds over Solar Energy?"

55. Patrick J. Michaels, "The Lessons of Solyndra: Green Swans, Opportunity Cost and Fast Neutrinos," *Forbes*, September 28, 2011. www.forbes.com.

56. Ronald Brownstein, "The China Energy Paradox," *National Journal*, June 4, 2010. www.nationaljournal.com.

57. Center for Climate and Energy Solutions, "Solar Power," August 2011. www.c2es.org.

58. Babinet et al., "Solar's Push to Reach the Mainstream."

59. Becky Harris, "The Future of Solar Cars," *Solar Power Blog*. http://thesolarpoweredproducts.com.

60. David Rotman, "Chasing the Sun," *Technology Review*, July/August 2009, p. 49.

Chapter Four: Should Government Play a Role in Developing Solar Power?

61. Solar Energy Industries Association, "Behind the Solyndra Headlines: America's Solar Energy Boom," September 19, 2011. www.seia.org.

62. Barack Obama, "State of the Union Address," Washington, DC, January 25, 2012.

63. Quoted in Rotman, "Chasing the Sun," p. 51.

64. Quoted in Christopher Joyce, "Cheap Chinese Panels Spark Solar Power Trade War," NPR, January 19, 2012. www.npr.org.

65. Quoted in Jason Margolis, "China's Grip on Solar Power," *PRI's The World*, October 6, 2011. www.theworld.org.

66. Quoted in Margolis, "China's Grip on Solar Power."

67. Richard W. Caperton, "Good Government Investments in Renewable Energy," Center for American Progress, January 10, 2012. www.americanprogress.org.

68. Caperton, "Good Government Investments in Renewable Energy."

69. Evan I. Schwartz, "The German Experiment," *Technology Review*, July/August 2010. www.technologyreview.com.

70. Solar Energy Industries Association, "Solar Energy Facts."

71. Nicolas Loris, "Competition, Not Handouts, Should Determine Role of Green Energy," *U.S. News & World Report*, January 18, 2012. www.usnews.com.

72. David Milroy, "Looking to Market Forces, Not Government, to Make the World Greener," *Huffington Post*, January 4, 2011. www.huffingtonpost.com.

73. William O'Keefe, "Let Markets Shape Our Energy Future," September 26, 2011, response to Amy Harder, "What Role Should Government Play in Energy Production?," *Energy Expert Blog, National Journal*, September 26, 2011. www.nationaljournal.com.

74. O'Keefe, "Let Markets Shape Our Energy Future."

75. Quoted in Eric Lipton and Clifford Krauss, "A Gold Rush of Subsidies in the Search for Clean Energy," *New York Times*, November 11, 2011. www.nytimes.com.

76. *Economist*, "Solar Power: Thou Orb Aloft Full-Dazzling," October 15, 2011. www.economist.com.

77. Dan Walters, "Cost of Reaching for the Sun Will Soar," *Sacramento Bee*, December 21, 2011. www.sacbee.com.

78. Patrick J. Michaels, "The Great Green Energy Crack-Up," *Forbes*, October 21, 2011. www.forbes.com.

79. Quoted in Peter Müller and Alexander Neubacher, "Solar Subsidy 'Insanity' Will Cost Consumers," *Der Spiegel*, January 16, 2012. www.spiegel.de.

80. Barack Obama, "Remarks by the President on the Economy," White House, May 26, 2010. www.whitehouse.gov.

81. Jonathan W. Emord, "The Lost Solyndra Lesson," *USA Today*, November 2011, p. 17.

82. Quoted in Erin Kelly, "Future of Federal Solar Programs in Doubt," *USA Today*, June 28, 2011. www.usatoday.com.

Solar Power Facts

Solar Power Around the World

- According to the Renewable Energy Policy Network for the 21st Century, in 2010 the world leaders in solar PV power were Germany, with 44 percent of the world's capacity; Spain, with 10 percent; and Japan and Italy, with 9 percent each.
- According to the Solar Energy Industries Association, solar technologies available today in the United States provide enough electricity to power 630,000 homes.
- The Center for Climate and Energy Solutions reports that from 2000 to 2010, the world's total solar energy capacity averaged 40 percent growth each year.
- The European Photovoltaic Industry Association says that of the 27.7 GW of new PV power added worldwide in 2011, almost 21 GW was added in Europe.
- According to the German Association of Energy and Water Industries, between 2010 and 2011 the number of PV installations in Germany increased 76 percent.

The Cost of Using Solar Power

- In 2011 the Solar Company estimates that a 3KW solar system that could power a typical California residence would have an estimated net cost of about $15,000, after all rebates and tax credits.
- According to the Center for Climate and Energy Solutions, the levelized cost—an average estimated cost over the lifetime of a power plant—of PV solar technology ranges from about 13.5 to 21.4 cents per kilowatt-hour of electricity produced. Costs for concentrated solar thermal plants are estimated to be 19.5 to 22.6 cents per kilowatt-hour.
- The DOE states that between 2009 and 2011, the average installed cost of residential and commercial PV solar systems fell at least 28 percent.

- The European Academies Science Advisory Council expects that over the next 15 to 20 years, the price of CSP generation will drop by 50 to 60 percent.

Solar Power and the Economy

- According to the Solar Foundation's National Solar Jobs Census, between 2009 and 2010 the number of people employed in the US solar industry doubled.
- According to the Solar Energy Industries Association, in 2011 the solar industry employed more than 100,000 Americans working at more than 5,000 companies.
- Data from research company Bloomberg New Energy Finance shows that investments in solar power in 2011 in the United States were $136.6 billion.
- The Solar Energy Industries Association reports that in 2010 the United States exported more solar products than it imported.

Public Opinion

- According to a 2011 survey of 1,000 US adults by Pike Research, 79 percent of respondents are in favor of using solar energy.
- In a 2011 survey of more than 100 senior executives in the US and Canadian natural gas and electric industries, research companies Capgemini and Platts found that a majority of the utilities plan to increase their use of solar power in the future.

Environmental Impact

- In a 2009 study by Robert McDonald and other researchers from the Nature Conservancy, the authors calculated the area of land needed to get 1 terawatt-hour of energy—about the same amount produced in a year by a small power plant—and found that for coal it is 3.75 square miles (9.7 sq km), concentrating solar thermal technology takes 5.9 square miles (15.3 sq km), and solar PV takes 14.25 square miles (36.9 sq km).

- According to SunPower, the developer of the California Valley Solar Ranch, the facility will power over 100,000 homes and will reduce carbon dioxide emissions by more than 750 million pounds (340 million kg) per year, the equivalent of removing 62,000 cars from local roads and freeways every year.
- According to the Solar Energy Industries Association, solar power uses far less water than some other industries. It says that agriculture in Nevada uses nearly four times as much water for the same area of land as the Nevada Solar One CSP plant in that state.

Future Potential

- In a 2010 report the International Energy Agency projects that by 2050, 11 percent of the world's electricity supplies will come from solar panels on homes and offices.
- According to the EIA, most of the world's electricity could be supplied by covering 4 percent of the world's desert area with PV installations.
- The International Energy Agency projects that PV solar power will provide 5 percent of global electricity consumption in 2030, rising to 11 percent in 2050.
- A study by Greenpeace International, the European Solar Thermal Electricity Association, and the International Energy Agency's SolarPACES group finds that by 2050, concentrated solar power could provide 25 percent of the world's energy needs.

Related Organizations and Websites

American Solar Energy Society (ASES)
4760 Walnut St., Suite 106
Boulder, CO 80301
phone: (303) 443-3130 • fax: (303) 443-3212
e-mail: ases@ases.org • website: www.ases.org

The ASES was established in 1954. It is a nonprofit association of solar professionals and advocates that works to promote energy innovation and a transition to a sustainable energy economy. The association publishes *Solar Today* magazine.

Center for Climate and Energy Solutions
2101 Wilson Blvd., Suite 550
Arlington, VA 22201
phone: (703) 516-4146 • fax: (703) 516-9551
website: www.c2es.org

The Center for Climate and Energy Solutions is a nonprofit organization that aims to address the challenges of climate change and energy. It promotes solar power as one of the solutions to the challenge of providing reliable and affordable energy for all, while protecting the global climate. The center's website has information about solar power and other types of renewable energy.

Citizens' Alliance for Responsible Energy (CARE)
PO Box 52103
Albuquerque, NM 87181
phone: (505) 239-8998
e-mail: info@responsiblenergy.org • website: www.responsiblenergy.org

CARE is an organization that advocates nuclear power and coal as more realistic energy sources. It argues that solar will never be enough to meet the world's growing energy demand.

International Solar Energy Society (ISES)
Villa Tannheim
Wiesentalstr. 50
79115 Freiburg
Germany
phone: 49-761-45906-0 • fax: 49-761-45906-99
e-mail: hq@ises.org • website: www.ises.org

The ISES has members in more than 50 countries. It works to advance renewable energy technology, including solar power, and to promote global cooperation in developing and implementing that technology. The ISES publishes several publications, including *Renewable Energy Focus*, the scientific journal *Solar Energy*, and various white papers.

National Renewable Energy Laboratory (NREL)
1617 Cole Blvd.
Golden, CO 80401
phone: (303) 275-3000
website: www.nrel.gov

The NREL is the DOE's laboratory for renewable energy research and development. Its website has maps, graphs, charts, and reports about renewable energy, including solar power.

Renewable Energy Policy Network for the 21st Century (REN21)
15 rue de Milan
75441 Paris Cedex 9 France
phone: 33 1 44 37 50 90 • fax: 33 1 44 37 50 95
e-mail: secretariat@ren21.net • website: www.ren21.net

REN21 is an organization that works to accelerate the use of renewable energy around the world. It promotes cooperation and the sharing of ideas and information, and works to advance policies that will increase the use of renewable energy.

Solar Done Right

e-mail: contact@solardoneright.org • website: http//:solardoneright.org

Solar Done Right is a coalition of solar power experts, biologists, activists, and others who believe that current solar power development threatens the environment. It advocates solar development in a way that is less harmful to the environment. The organization's website contains fact sheets and briefings about solar power.

Solar Electric Power Association (SEPA)

1220 Nineteenth St. NW, Suite 800
Washington, DC 20036
phone: (202) 857-0898
website: www.solarelectricpower.org

The SEPA is a nonprofit organization dedicated to helping utilities integrate solar power into their portfolios. It believes this can benefit both utilities and power consumers. Its website has solar information basics and cost estimates.

Solar Energy Industries Association (SEIA)

phone: (202) 682-0556
e-mail: info@seia.org • website: www.seia.org

The SEIA works to promote the use of solar energy in the United States. Its website contains research about solar power, information on current solar issues, and numerous fact sheets.

Solar Foundation

575 Seventh St. NW, Suite 400
Washington, DC 20004
phone: (202) 469-3750
website: www.thesolarfoundation.org

The Solar Foundation is a nonprofit organization that works to demonstrate the global benefits of solar energy. It hopes to increase the adoption of solar power through research and education. Its website contains research and a blog about various solar power issues.

For Further Research

Books

Daniel D. Chiras, Robert Aram, and Kurt Nelson, *Solar Electricity Basics*. Gabriola Island, BC: New Society, 2010.

Robert Foster, *Solar Energy: Renewable Energy and the Environment*. Boca Raton, FL: CRC, 2010.

Richard Hantula, *How Do Solar Panels Work?* New York: Chelsea Clubhouse, 2010.

Alireza Khaligh, *Energy Harvesting: Solar, Wind, and Ocean Energy Conversion Systems*. Boca Raton, FL: Taylor Francis, 2010.

Paul A. Lynn, *Electricity from Sunlight: An Introduction to Photovoltaics*. Chichester, UK: Wiley, 2010.

Wolfgang W. Palz, *Power for the World: The Emergence of Electricity from the Sun*. Singapore: Pan Stanford, 2011.

Periodicals

The Economist, "Solar Power: A Painful Eclipse" October 15, 2011.

James R. Fischer and Gale Buchanan, "Developing Solar Power," *Resource*, May/June 2010.

Robert Glennon, "Storm Clouds over Solar Energy?," *Solar Today*, April 2011.

Monique Hanis, "To Fight Big Pollution, We Need Big Solar," *Earth Island Journal*, Autumn 2010.

Steven Mufson, "Before Solyndra, a Long History of Failed Government Energy Projects," *Washington Post*, November 11, 2011.

Evan I. Schwartz, "The German Experiment," *Technology Review*, July/August 2010.

Internet Sources

Amy Harder, "What Role Should Government Play in Energy Production?," *Energy Experts Blog, National Journal*, September 26, 2011. http://energy.nationaljournal.com/2011/09/what-role-should-government-pl.php.

National Renewable Energy Laboratory, "Solar Power and the Electric Grid," March 2010. www.nrel.gov/csp/pdfs/45653.pdf.

Pew Center on Global Climate Change, "Solar Power," Climate Techbook, August 2011. www.c2es.org/technology/factsheet/solar.

Solar Energy Industries Association, "Solar Energy Facts: Correcting Old Myths," September 23, 2011. www.seia.org/galleries/pdf/Solar_Energy_Facts_-_Correcting_Old_Myths.pdf.

Solar Foundation, "National Solar Jobs Census 2010: A Review of the U.S. Solar Workforce," October 2010. www.thesolarfoundation.org/sites/thesolarfoundation.org/files/Final%20TSF%20National%20Solar%20Jobs%20Census%202010%20Web%20Version.pdf.

Index